DISCRETE MATHEMATICS AND ITS APPLICATIONS

Spanning Trees and Optimization Problems

DISCRETE MATHEMATICS AND ITS APPLICATIONS

Series Editor
Kenneth H. Rosen, Ph.D.
AT&T Laboratories
Middletown, New Jersey

Charles J. Colbourn and Jeffrey H. Dinitz, The CRC Handbook of Combinatorial Designs

Charalambos A. Charalambides, Enumerative Combinatorics

Steven Furino, Ying Miao, and Jianxing Yin, Frames and Resolvable Designs: Uses, Constructions, and Existence

Randy Goldberg and Lance Riek, A Practical Handbook of Speech Coders

Jacob E. Goodman and Joseph O'Rourke, Handbook of Discrete and Computational Geometry

Jonathan Gross and Jay Yellen, Graph Theory and Its Applications

Jonathan Gross and Jay Yellen, Handbook of Graph Theory

Darrel R. Hankerson, Greg A. Harris, and Peter D. Johnson, Introduction to Information Theory and Data Compression

Daryl D. Harms, Miroslav Kraetzl, Charles J. Colbourn, and John S. Devitt, Network Reliability: Experiments with a Symbolic Algebra Environment

David M. Jackson and Terry I. Visentin, An Atlas of Smaller Maps in Orientable and Nonorientable Surfaces

Richard E. Klima, Ernest Stitzinger, and Neil P. Sigmon, Abstract Algebra Applications with Maple

Patrick Knupp and Kambiz Salari, Verification of Computer Codes in Computational Science and Engineering

Donald L. Kreher and Douglas R. Stinson, Combinatorial Algorithms: Generation Enumeration and Search

Charles C. Lindner and Christopher A. Rodgers, Design Theory

Alfred J. Menezes, Paul C. van Oorschot, and Scott A. Vanstone, Handbook of Applied Cryptography

Richard A. Mollin, Algebraic Number Theory

Richard A. Mollin, Fundamental Number Theory with Applications

Richard A. Mollin, An Introduction to Cryptography

Richard A. Mollin, Quadratics

Continued Titles

Richard A. Mollin, RSA and Public-Key Cryptography

Kenneth H. Rosen, Handbook of Discrete and Combinatorial Mathematics

Douglas R. Shier and K.T. Wallenius, Applied Mathematical Modeling: A Multidisciplinary Approach

Douglas R. Stinson, Cryptography: Theory and Practice, Second Edition

Roberto Togneri and Christopher J. deSilva, Fundamentals of Information Theory and Coding Design

Lawrence C. Washington, Elliptic Curves: Number Theory and Cryptography

Bang Ye Wu and Kun-Mao Chao, Spanning Trees and Optimization Problems

DISCRETE MATHEMATICS AND ITS APPLICATIONS

Spanning Trees and Optimization Problems

Bang Ye Wu
Department of Computer Science and Information Engineering, Shu-Te University, Yen-Chau, Kaohsiung County, Taiwan 824

Kun-Mao Chao
Department of Computer Science and Information Engineering, National Taiwan University, Taipei, Taiwan 106

CHAPMAN & HALL/CRC

A CRC Press Company
Boca Raton   London   New York   Washington, D.C.

Library of Congress Cataloging-in-Publication Data

Wu, Bang Ye.
  Spanning trees and optimization problems / Bang Ye Wu and Kun-Mao Chao.
    p. cm. — (Discrete mathematics and its applications)
  Includes bibliographical references and index.
  ISBN 1-58488-436-3 (alk. paper)
    1. Trees (Graph theory)—Data processing. 2. Mathematical optimization. I. Chao, Kun-Mao. II. Title. III. Series.

QA166.2.W8 2004
511′.52—dc22   2003063533

This book contains information obtained from authentic and highly regarded sources. Reprinted material is quoted with permission, and sources are indicated. A wide variety of references are listed. Reasonable efforts have been made to publish reliable data and information, but the author and the publisher cannot assume responsibility for the validity of all materials or for the consequences of their use.

Neither this book nor any part may be reproduced or transmitted in any form or by any means, electronic or mechanical, including photocopying, microfilming, and recording, or by any information storage or retrieval system, without prior permission in writing from the publisher.

The consent of CRC Press LLC does not extend to copying for general distribution, for promotion, for creating new works, or for resale. Specific permission must be obtained in writing from CRC Press LLC for such copying.

Direct all inquiries to CRC Press LLC, 2000 N.W. Corporate Blvd., Boca Raton, Florida 33431.

Trademark Notice: Product or corporate names may be trademarks or registered trademarks, and are used only for identification and explanation, without intent to infringe.

Visit the CRC Press Web site at www.crcpress.com

© 2004 by Chapman & Hall/CRC

No claim to original U.S. Government works
International Standard Book Number 1-58488-436-3
Library of Congress Card Number 2003063533
Printed in the United States of America 1 2 3 4 5 6 7 8 9 0
Printed on acid-free paper

Preface

The research on spanning trees has been one of the most important areas in algorithm design. People who are interested in algorithms will find this book informative and inspiring. The new results are still accumulating, and we try to make clear the whole picture of the current status and future developments.

This book is written for graduate or advanced undergraduate students in computer science, electrical engineering, industrial engineering, and mathematics. It is also a good reference for professionals.

Our motivations for writing this book:

1. To the best of our knowledge, there is no book totally dedicated to the topics of spanning trees.

2. Our recent progress in spanning trees reveals a new line of investigation.

3. Designing approximation algorithms for spanning tree problems has become an exciting and important field in theoretical computer science.

4. Besides numerous network design applications, spanning trees have also been playing important roles in newly established research areas, such as biological sequence alignments, and evolutionary tree construction.

This book is a general and rigorous text on algorithms for spanning trees. It covers the full spectrum of spanning tree algorithms from classical computer science to modern applications. The selected topics in this book make it an excellent handbook on algorithms for spanning trees. At the end of every chapter, we report related work and recent progress.

We first explain general properties of spanning trees. We then focus on three categories of spanning trees, namely, minimum spanning trees, shortest-paths trees, and optimum routing cost spanning trees. We also show how to balance the tree costs. Besides the theoretical description of the methods, many examples are used to illustrate the ideas behind them. Moreover, we demonstrate some applications of these spanning trees. We explore in details some other interesting spanning trees, including maximum leaf spanning trees and minimum diameter spanning trees. In addition, Steiner trees and evolutionary trees are also discussed. We close this book by summarizing other important problems related to spanning trees.

Writing a book is not as easy as we thought at the very beginning of this project. We have tried our best to make it consistent and correct. However, it's a mission impossible for imperfect authors to produce a perfect book.

Should you find any mathematical, historical, or typographical errors, please let us know.

We are extremely grateful to Richard Chia-Tung Lee, Webb Miller, and Chuan Yi Tang, who always make the subject of algorithms exciting and beautiful in their superb lectures. Their guidance and suggestions throughout this study were indispensable.

We thank Vineet Bafna, Giuseppe Lancia, and R. Ravi for their collaborations on the minimum routing cost spanning tree problem. We thank Tao Jiang and Howard Karloff for suggesting the merger of two different works at the early stage of our investigation. We are also thankful to Piotr Berman, Xiaoqiu Huang, Yuh-Dauh Lyuu, Anna Östlin, Pavel Pevzner, and the anonymous reviewers for their valuable comments.

It has been a pleasure working with CRC Press in the development of this book. We are very proud to have this book included in the CRC series on Discrete Mathematics and Its Applications, edited by Kenneth H. Rosen. Ken also provided critical reviews and invaluable information for which we are grateful. We thank Sunil Nair for his final approval of our proposal. Richard O'Hanley was the first to approach us about the possibility of publishing a book at CRC Press. Robert B. Stern then handled the proposal review and contract arrangements efficiently. Bob also proposed many constructive suggestions throughout the project. Jamie B. Sigal helped us with both production and permissions issues, and his gentle reminders kept us moving at a good pace. William R. Palmer III resolved the questions arising in prepress. Nishith Arora revised the LaTeX style files in a timely manner. Julie Spadaro set the production schedule in a perfect way, and kindly copyedited our manuscript for us to review and correct.

Finally, we thank our families for their love, patience, and encouragement. We thank our wives, Mei-Ling Cheng and Pei-Ju Tsai, and our sons, Ming-Hsuan Wu and Leo Liang Chao, for tolerating our absentmindedness during the writing of this book. We promise to work less than 168 hours a week by not taking on a new grand project immediately.

Bang Ye Wu
bangye@mail.stu.edu.tw
http://www.personal.stu.edu.tw/bangye

Kun-Mao Chao
kmchao@csie.ntu.edu.tw
http://www.csie.ntu.edu.tw/~kmchao

December 2003

About the Authors

Bang Ye Wu was born in Kaohsiung, Taiwan, in 1964. He earned the B.S. degree in electrical engineering from Chung Cheng Institute of Technology, Taiwan, in 1986, and the M.S. and the Ph.D. degrees in computer science from National Tsing-Hua University, Taiwan, in 1991 and 1999 respectively. He is currently an assistant professor and the head of the Department of Computer Science and Information Engineering, Shu-Te University. Before joining the faculty of Shu-Te University, he worked in the Chung-Shan Institute of Science and Technology, Taiwan, as a research assistant (1986-1989), an assistant research fellow (1991-1995), and an associate research fellow (1999-2000). His current research interests include algorithms and bioinformatics.

Kun-Mao Chao was born in Tou-Liu, Taiwan, in 1963. He earned the B.S. and M.S. degrees in computer engineering from National Chiao-Tung University, Taiwan, in 1985 and 1987, respectively, and the Ph.D. degree in computer science from The Pennsylvania State University, University Park, in 1993. He is currently a professor of the Department of Computer Science and Information Engineering, National Taiwan University. From 1987 to 1989, he served in the ROC Air Force Headquarters as a system engineer. From 1993 to 1994, he worked as a postdoctoral fellow at Penn State's Center for Computational Biology. In 1994, he was a visiting research scientist at the National Center for Biotechnology Information, National Institutes of Health, Bethesda, Maryland. Before joining the faculty of National Taiwan University, he taught in the Department of Computer Science and Information Management, Providence University, from 1994 to 1999, and the Department of Life Science, National Yang-Ming University, from 1999 to 2002. His current research interests include algorithms and bioinformatics. Dr. Chao is a member of Phi Tau Phi and Phi Kappa Phi.

Contents

1 Spanning Trees   1
    1.1 Counting Spanning Trees   1

2 Minimum Spanning Trees   9
    2.1 Introduction   9
    2.2 Borůvka's Algorithm   11
    2.3 Prim's Algorithm   13
    2.4 Kruskal's Algorithm   15
    2.5 Applications   17
       2.5.1 Cable TV   17
       2.5.2 Circuit design   17
       2.5.3 Islands connection   17
       2.5.4 Clustering gene expression data   17
       2.5.5 MST-based approximations   18
    2.6 Summary   18
    Bibliographic Notes and Further Reading   19
    Exercises   20

3 Shortest-Paths Trees   23
    3.1 Introduction   23
    3.2 Dijkstra's Algorithm   25
    3.3 The Bellman-Ford Algorithm   33
    3.4 Applications   35
       3.4.1 Multicast   37
       3.4.2 SPT-based approximations   37
    3.5 Summary   38
    Bibliographic Notes and Further Reading   38
    Exercises   39

4 Minimum Routing Cost Spanning Trees   41
    4.1 Introduction   41
    4.2 Approximating by a Shortest-Paths Tree   44
       4.2.1 A simple analysis   44
       4.2.2 Solution decomposition   46
    4.3 Approximating by a General Star   47
       4.3.1 Separators and general stars   47
       4.3.2 A 15/8-approximation algorithm   52

	4.3.3	A 3/2-approximation algorithm	55
	4.3.4	Further improvement	57
4.4	A Reduction to the Metric Case		58
4.5	A Polynomial Time Approximation Scheme		62
	4.5.1	Overview	62
	4.5.2	The δ-spine of a tree	66
	4.5.3	Lower bound	69
	4.5.4	From trees to stars	70
	4.5.5	Finding an optimal k-star	74
4.6	Applications		79
	4.6.1	Network design	79
	4.6.2	Computational biology	79
4.7	Summary		82
Bibliographic Notes and Further Reading			82
Exercises			83

5 Optimal Communication Spanning Trees — 85

- 5.1 Introduction — 85
- 5.2 Product-Requirement — 87
  - 5.2.1 Overview — 87
  - 5.2.2 Preliminaries — 88
  - 5.2.3 Approximating by 2-stars — 91
  - 5.2.4 A polynomial time approximation scheme — 98
- 5.3 Sum-Requirement — 104
- 5.4 Multiple Sources — 109
  - 5.4.1 Computational complexity for fixed p — 110
  - 5.4.2 A PTAS for the 2-MRCT — 115
- 5.5 Applications — 124
- 5.6 Summary — 125
- Bibliographic Notes and Further Reading — 125
- Exercises — 127

6 Balancing the Tree Costs — 129

- 6.1 Introduction — 129
- 6.2 Light Approximate Shortest-Paths Trees — 130
  - 6.2.1 Overview — 130
  - 6.2.2 The algorithm — 131
  - 6.2.3 The analysis of the algorithm — 134
- 6.3 Light Approximate Routing Cost Spanning Trees — 136
  - 6.3.1 Overview — 136
  - 6.3.2 The algorithm — 137
  - 6.3.3 The performance analysis — 140
  - 6.3.4 On general graphs — 143
- 6.4 Applications — 143
- 6.5 Summary — 144

	Bibliographic Notes and Further Reading	144
	Exercises	145
7	**Steiner Trees and Some Other Problems**	**147**
	7.1 Steiner Minimal Trees	147
	7.1.1 Approximation by MST	148
	7.1.2 Improved approximation algorithms	151
	7.2 Trees and Diameters	154
	7.2.1 Eccentricities, diameters, and radii	154
	7.2.2 The minimum diameter spanning trees	157
	7.3 Maximum Leaf Spanning Trees	162
	7.3.1 Leafy trees and leafy forests	162
	7.3.2 The algorithm	165
	7.3.3 Performance ratio	166
	7.4 Some Other Problems	168
	7.4.1 Network design	169
	7.4.2 Computational biology	170
	Bibliographic Notes and Further Reading	173
	Exercises	174

References   **175**

Index   **183**

Symbol Description

$d_G(u,v)$	the distance (shortest path length) between u and v on G, i.e., $w(SP_G(u,v))$
$d_G(v,U)$	the minimum distance from vertex v to any vertex in U, i.e., $\min_{u \in U}\{d_G(v,u)\}$
$D_G(v,U)$	the maximum distance from vertex v to any vertex in U, i.e., $\max_{u \in U}\{d_G(v,u)\}$
$E(G)$	the edge set of graph G
$G = (V, E, w)$	a graph G with vertex set V, edge set E, and edge weight function $w : E \to Z_0^+$
$\bar{G}$	the metric closure of a graph G
$G - e$	the graph obtained by removing edge e from G
$G + e$	the graph obtained by inserting edge e into G
$G \backslash e$	the resulting graph after contracting e in G
m	the number of edges of the input graph
n	the number of vertices of the input graph
$parent(v)$	the parent of a vertex v in a rooted tree
$|S|$	the cardinality of a set S
$SP_G(u,v)$	a shortest path between vertices u and v on graph G
$V(G)$	the vertex set of graph G
$w(G)$	the total length of edges in $E(G)$, i.e., $\sum_{e \in E(G)} w(e)$
Z^+	the set of all positive integers
Z_0^+	the set of all nonnegative integers

Chapter 1

Spanning Trees

This book provides a comprehensive introduction to the modern study of spanning trees. A spanning tree for a graph G is a subgraph of G that is a tree and contains all the vertices of G. There are many situations in which good spanning trees must be found. Whenever one wants to find a simple, cheap, yet efficient way to connect a set of terminals, be they computers, telephones, factories, or cities, a solution is normally one kind of spanning trees. Spanning trees prove important for several reasons:

1. They create a sparse subgraph that reflects a lot about the original graph.

2. They play an important role in designing efficient routing algorithms.

3. Some computationally hard problems, such as the Steiner tree problem and the traveling salesperson problem, can be solved approximately by using spanning trees.

4. They have wide applications in many areas, such as network design, bioinformatics, etc.

1.1 Counting Spanning Trees

Throughout this book, we use n to denote the number of vertices of the input graph, and m the number of edges of the input graph. Let us start with the problem of counting the number of spanning trees. Let K_n denote a complete graph with n vertices. How many spanning trees are there in the complete graph K_n? Before answering this question, consider the following simpler question. How many trees are there spanning all the vertices in Figure 1.1?

The answer is 16. Figure 1.2 gives all 16 spanning trees of the four-vertex complete graph in Figure 1.1. Each spanning tree is associated with a two-number sequence, called a Prüfer sequence, which will be explained later.

Back in 1889, Cayley devised the well-known formula n^{n-2} for the number of spanning trees in the complete graph K_n [17]. There are numerous proofs

1

FIGURE 1.1: A four-vertex complete graph K_4.

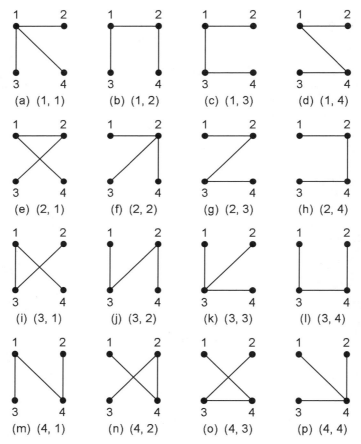

FIGURE 1.2: All 16 spanning trees of K_4.

of this elegant formula. The first explicit combinatorial proof of Cayley's formula is due to Prüfer [80]. The idea of Prüfer's proof is to find a one-to-one correspondence (bijection) between the set of spanning trees of K_n, and the set of Prüfer sequences of length $n-2$, which is defined in Definition 1.1.

DEFINITION 1.1 A Prüfer sequence of length $n-2$, for $n \geq 2$, is any sequence of integers between 1 and n, with repetitions allowed.

LEMMA 1.1
There are n^{n-2} Prüfer sequences of length $n-2$.

PROOF By definition, there are n ways to choose each element of a Prüfer sequence of length $n-2$. Since there are $n-2$ elements to be determined, in total we have n^{n-2} ways to choose the whole sequence. □

Example 1.1
The set of Prüfer sequences of length 2 is $\{(1,1), (1,2), (1,3), (1,4), (2,1),$ $(2,2), (2,3), (2,4), (3,1), (3,2), (3,3), (3,4), (4,1), (4,2), (4,3), (4,4)\}$. In total, there are $4^{4-2} = 16$ Prüfer sequences of length 2. □

Given a labeled tree with vertices labeled by $1, 2, 3, \ldots, n$, the PRÜFER ENCODING algorithm outputs a unique Prüfer sequence of length $n-2$. It initializes with an empty sequence. If the tree has more than two vertices, the algorithm finds the leaf with the lowest label, and appends to the sequence the label of the neighbor of that leaf. Then the leaf with the lowest label is deleted from the tree. This operation is repeated $n-2$ times until only two vertices remain in the tree. The algorithm ends up deleting $n-2$ vertices. Therefore, the resulting sequence is of length $n-2$.

Algorithm: PRÜFER ENCODING
Input: A labeled tree with vertices labeled by $1, 2, 3, \ldots, n$.
Output: A Prüfer sequence.
    Repeat $n-2$ times
        $v \leftarrow$ the leaf with the lowest label
        Put the label of v's unique neighbor in the output sequence.
        Remove v from the tree.

Let us look at Figure 1.2 once again. In Figure 1.2(a), vertex 2 is the leaf with the lowest label, thus we add its unique neighbor vertex 1 to the sequence. After removing vertex 2 from the tree, vertex 3 becomes the leaf with the lowest label, and we again add its unique neighbor vertex 1 to the sequence. Therefore, the resulting Prüfer sequence is $(1,1)$. In Figure 1.2(b), vertex 3 is the leaf with the lowest label, thus we add its unique neighbor vertex 1 to the sequence. After removing vertex 3 from the tree, vertex 1 becomes the leaf with the lowest label, and we add its unique neighbor vertex 2 to the sequence. Therefore, the resulting Prüfer sequence is $(1,2)$.

Now consider a more complicated tree in Figure 1.3. What is its corresponding Prüfer sequence?

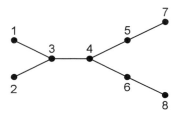

FIGURE 1.3: An eight-vertex spanning tree.

Figure 1.4 illustrates the encoding process step by step. In Figure 1.4(a), vertex 1 is the leaf with the lowest label, thus we add its unique neighbor vertex 3 to the sequence, resulting in the Prüfer sequence under construction, denoted by P, equals to (3). Then we remove vertex 1 from the tree. In Figure 1.4(b), vertex 2 becomes the leaf with the lowest label; we again add its unique neighbor vertex 3 to the sequence. So we have $P = (3,3)$. After removing vertex 2 from the tree, vertex 3 is the leaf with the lowest label. Since vertex 4 is the unique label of vertex, we get $P = (3,3,4)$. Repeat this operation a few times until only two vertices remain in the tree. In this example, vertices 6 and 8 are the two vertices left. It follows that the Prüfer sequence for Figure 1.3 is $(3,3,4,5,4,6)$.

It can be verified that different spanning trees of K_n determine different Prüfer sequences. The PRÜFER DECODING algorithm provides the inverse algorithm, which finds the unique labeled tree T with n vertices for a given Prüfer sequence of $n-2$ elements. Let the given Prüfer sequence be $P = (p_1, p_2, \ldots, p_{n-2})$. Observe that any vertex v of T occurs $\deg(v) - 1$ times in $(p_1, p_2, \ldots, p_{n-2})$, where $\deg(v)$ is the degree of vertex v. Thus the vertices of degree one, i.e., the leaves, in T are those that do not appear in P. To reconstruct T from $(p_1, p_2, \ldots, p_{n-2})$, we proceed as follows. Let V be the vertex label set $\{1, 2, \ldots, n\}$. In the ith iteration of the **for** loop, $P = (p_i, p_{i+1}, \ldots, p_{n-2})$. Let v be the smallest element of the set V that does not occur in P. We connect vertex v to vertex p_i. Then we remove v from V, and p_i from P. Repeat this process for $n-2$ times until only two numbers are left in V. Finally, we connect the vertices corresponding to the remaining two numbers in V. It can be shown that different Prüfer sequences deliver different spanning trees of K_n.

Algorithm: PRÜFER DECODING
Input: A Prüfer sequence $P = (p_1, p_2, \ldots, p_{n-2})$.
Output: A labeled tree with vertices labeled by $1, 2, 3, \ldots, n$.
    $P \leftarrow$ the input Prüfer sequence
    $n \leftarrow |P| + 2$
    $V \leftarrow \{1, 2, \ldots, n\}$

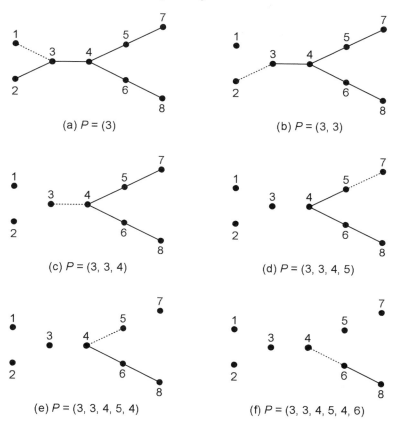

FIGURE 1.4: Generating a Prüfer sequence from a spanning tree.

Start with n isolated vertices labeled $1, 2, \ldots, n$.
for $i = 1$ **to** $n - 2$ **do**
    $v \leftarrow$ the smallest element of the set V that does not occur in P
    Connect vertex v to vertex p_i
    Remove v from the set V
    Remove the element p_i from the sequence P
    /* Now $P = (p_{i+1}, p_{i+2}, \ldots, p_{n-2})$ */
Connect the vertices corresponding to the two numbers in V.

To see how the PRÜFER DECODING algorithm works, let us build the spanning tree corresponding to the Prüfer sequence $(3, 3, 4, 5, 4, 6)$. Figure 1.5 illustrates the decoding process step by step. Initially, $P = (3, 3, 4, 5, 4, 6)$ and $V = \{1, 2, 3, 4, 5, 6, 7\}$. Vertex 1 is the smallest element of the set V that does not occur in P. Thus we connect vertex 1 to the first element of P, i.e., vertex 3 (see Figure 1.5(a)). Then we remove 3 from P, and 1 from V. Now $P = (3, 4, 5, 4, 6)$ and $V = \{2, 3, 4, 5, 6, 7, 8\}$. We connect vertex 2 to vertex 3. Repeat this operation again and again until only two numbers are left in V (see Figure 1.5(g)).

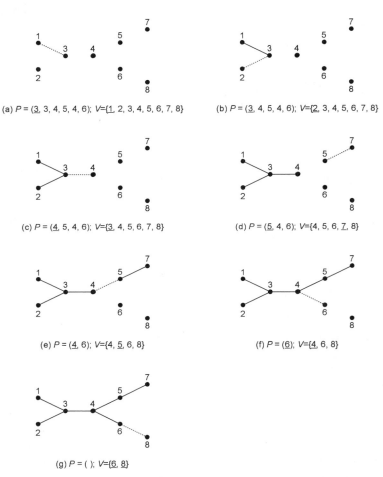

FIGURE 1.5: Recovering a spanning tree from a Prüfer sequence.

We have established a one-to-one correspondence (bijection) between the set of spanning trees of K_n, and the set of Prüfer sequences of length $n - 2$. We obtain the result in Theorem 1.1.

THEOREM 1.1
The number of spanning trees in K_n is n^{n-2}.

It should be noted that n^{n-2} is the number of distinct spanning trees of K_n, but not the number of nonisomorphic spanning trees of K_n. For example, there are $6^{6-2} = 1296$ distinct spanning trees of K_6, yet there are only six nonisomorphic spanning trees of K_6.

In the following, we give a recursive formula for the number of spanning trees in a general graph. Let $G - e$ denote the graph obtained by removing edge e from G. Let $G \backslash e$ denote the resulting graph after contracting e in G. In other words, $G \backslash e$ is the graph obtained by deleting e, and merging its ends. Let $\tau(G)$ denote the number of spanning trees of G. The following recursive formula computes the number of spanning trees in a graph.

THEOREM 1.2
$\tau(G) = \tau(G - e) + \tau(G \backslash e)$

PROOF The number of spanning trees of G that do not contain e is $\tau(G-e)$ since each of them is also a spanning tree of $G-e$, and vice versa. On the other hand, the number of spanning trees of G that contain e is $\tau(G \backslash e)$ because each of them corresponds to a spanning tree of $G \backslash e$. Therefore, $\tau(G) = \tau(G - e) + \tau(G \backslash e)$. □

Chapter 2

Minimum Spanning Trees

2.1 Introduction

Suppose you have a business with several branch offices and you want to lease phone lines to connect them with each other. Your goal is to connect all your offices with the minimum total cost. The resulting connection should be a spanning tree since if it is not a tree, you can always remove some edges without losing the connectivity to save money.

A minimum spanning tree (MST) of a weighted graph G is a spanning tree of G whose edges sum to minimum weight. In other words, a minimum spanning tree is a tree formed from a subset of the edges in a given undirected graph, with two properties: (1) it spans the graph, i.e., it includes every vertex in the graph, and (2) it is a minimum, i.e., the total weight of all the edges is as low as possible.

The minimum spanning tree problem is always included in algorithm textbooks since (1) it arises in many applications, (2) it is an important example where greedy algorithms always deliver an optimal solution, and (3) clever data structures are necessary to make it work efficiently.

What is a minimum spanning tree for the weighted graph in Figure 2.1? Notice that a minimum spanning tree is not necessarily unique.

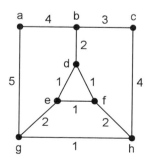

FIGURE 2.1: A weighted graph.

9

Figure 2.2 gives four minimum spanning trees, where each of them is of total weight 14. These trees can be derived by growing the spanning tree in a *greedy* way.

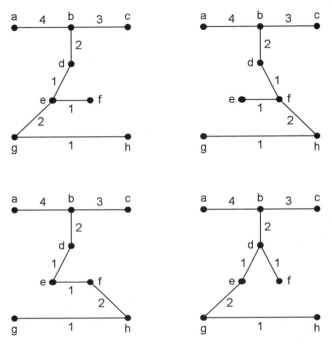

FIGURE 2.2: Some minimum spanning trees.

Before exploring the MST algorithms, we state some important facts about spanning trees. Let $G + e$ denote the graph obtained by inserting edge e into G.

LEMMA 2.1
Any two vertices in a tree are connected by a unique path.

PROOF Since a tree is connected, any two vertices in a tree are connected by at least one simple path. Let T be a tree, and assume that there are two distinct paths P_1 and P_2 from vertex u to vertex v. There exists an edge $e = (x, y)$ of P_1 that is not an edge of P_2. It can be seen that $(P_1 \cup P_2) - e$ is connected, and it contains a path P_{xy} from vertex x to vertex y. But then $P_{xy} + e$ is a cycle. This is a contradiction. Thus, there can be at most one path between two vertices in a tree. □

LEMMA 2.2
Let T be a spanning tree of a graph G, and let e be an edge of G not in T. Then $T + e$ contains a unique cycle.

PROOF Let $e = (u, v)$. Since T is acyclic, each cycle of $T + e$ contains e. Moreover, X is a cycle of $T + e$ if and only if $X - e$ is a path from u to v in T. By Lemma 2.1, such a path is unique in T. Thus $T + e$ contains a unique cycle. □

In this chapter, we shall examine three well-known algorithms for solving the minimum spanning tree problem: Borůvka's algorithm, Prim's algorithm, and Kruskal's algorithm. They all exploit the following fact in one way or another.

THEOREM 2.1
Let $F_1, F_2, \ldots, F_k$ be a spanning forest of G, and let (u, v) be the smallest of all edges with only one endpoint $u \in V(F_1)$. Then there is an optimal one containing (u, v) among all spanning trees containing all edges in $\cup_{i=1}^{k} E(F_i)$.

PROOF By contradiction. Suppose that there is a spanning tree T of G with $\cup_{i=1}^{k} E(F_i) \subseteq E(T)$, and $(u, v) \notin E(T)$, which is smaller than all spanning trees containing $\cup_{i=1}^{k} E(F_i) \cup \{(u, v)\}$. By Lemma 2.2, $T + (u, v)$ contains a unique cycle. Since $v \notin V(F_1)$, this cycle contains some vertices outside F_1. Thus there exists an edge (u', v'), different from (u, v), on this cycle such that $u' \in V(F_1)$ and $v' \notin V(F_1)$. Since this edge is no smaller than (u, v), and does not belong to $\cup_{i=1}^{k} E(F_i)$, $T + (u, v) - (u', v')$ is a new spanning tree with total weight no more than T. This is a contradiction. It follows that there is an optimal one containing (u, v) among all spanning trees containing all edges in $\cup_{i=1}^{k} E(F_i)$. □

2.2 Borůvka's Algorithm

The earliest known algorithm for finding a minimum spanning tree was given by Otakar Borůvka back in 1926. In a Borůvka step, every *supervertex* selects its smallest adjacent edge. These edges are added to the MST, avoiding cycles. Then the new supervertices, i.e., the connected components, are calculated by contracting the graph on the edges just added to the MST. This

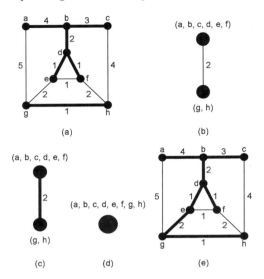

FIGURE 2.3: The execution of the BORŮVKA algorithm on the graph from Figure 2.1.

process is repeated until only one supervertex is left. In other words, there are $n-1$ edges contracted. The union of these edges gives rise to a minimum spanning tree.

Algorithm: BORŮVKA
Input: A weighted, undirected graph $G = (V, E, w)$.
Output: A minimum spanning tree T
    $T \leftarrow \emptyset$
    while $|T| < n-1$ **do**
        $F \leftarrow$ a forest consisting of the smallest edge incident to
            each vertex in G
        $G \leftarrow G \backslash F$
        $T \leftarrow T \cup F$

Figure 2.3 illustrates the execution of the BORŮVKA algorithm on the graph from Figure 2.1. In Figure 2.3(a), each vertex chooses the smallest incident edge without causing cycles. In Figure 2.3(b), vertices a, b, c, d, e, and f are contracted into one supervertex, and vertices g and h are contracted into the other supervertex. In Figure 2.3(d), these two supervertices are contracted into one supervertex. All the contracted edges constitute a minimum spanning tree as shown in Figure 2.3(e).

2.3 Prim's Algorithm

Prim's algorithm was conceived by computer scientist Robert Prim in 1957. It starts from an arbitrary vertex, and builds upon a single partial minimum spanning tree, at each step adding an edge connecting the vertex nearest to but not already in the current partial minimum spanning tree. It grows until the tree spans all the vertices in the input graph. This strategy is greedy in the sense that at each step the partial spanning tree is augmented with an edge that is the smallest among all possible neighboring edges.

Algorithm: PRIM
Input: A weighted, undirected graph $G = (V, E, w)$.
Output: A minimum spanning tree T.
    $T \leftarrow \emptyset$
    Let r be an arbitrarily chosen vertex from V.
    $U \leftarrow \{r\}$
    while $|U| < n$ **do**
        Find $u \in U$ and $v \in V - U$ such that the edge (u, v) is a smallest
            edge between U and $V - U$.
        $T \leftarrow T \cup \{(u, v)\}$
        $U \leftarrow U \cup \{v\}$

Figure 2.4 illustrates the execution of the PRIM algorithm on the graph from Figure 2.1. It starts at vertex a. Since (a, b) is the smallest edge incident to a, it is included in the spanning tree under construction (see Figure 2.4(a)). In Figure 2.4(b), (b, d) is added because it is the smallest edge between $\{a, b\}$ and $V - \{a, b\}$. When there is a tie, as in the situation in Figure 2.4(c), any smallest edge would work well. Proceed this way until all vertices are spanned. The final minimum spanning tree is shown in Figure 2.4(h).

Prim's algorithm appears to spend most of its time finding the smallest edge to grow. A straightforward method finds the smallest edge by searching the adjacency lists of the vertices in V; then each iteration costs $O(m)$ time, yielding a total running time of $O(mn)$. By using binary heaps, this can be improved to $O(m \log n)$. By using Fibonacci heaps, Prim's algorithm runs in $O(m + n \log n)$ time. Interested readers should refer to the end of this chapter for further improvements.

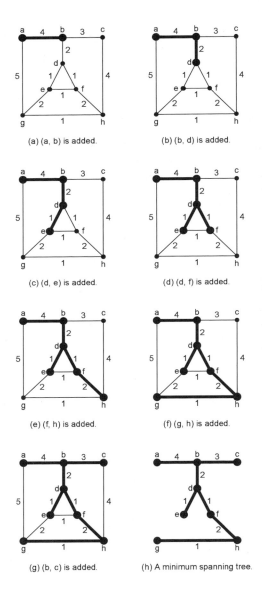

FIGURE 2.4: The execution of the PRIM algorithm on the graph from Figure 2.1.

2.4 Kruskal's Algorithm

Kruskal's algorithm was given by Joseph Kruskal in 1956. It creates a forest where each vertex in the graph is initially a separate tree. It then sorts all the edges in the graph. For each edge (u, v) in sorted order, we do the following. If vertices u and v belong to two different trees, then add (u, v) to the forest, combining two trees into a single tree. It proceeds until all the edges have been processed.

Algorithm: KRUSKAL
Input: A weighted, undirected graph $G = (V, E, w)$.
Output: A minimum spanning tree T.
    Sort the edges in E in nondecreasing order by weight.
    $T \leftarrow \emptyset$
    Create one set for each vertex.
    for each edge (u, v) in sorted order **do**
        $x \leftarrow$ FIND(u)
        $y \leftarrow$ FIND(v)
        if $x \neq y$ **then**
            $T \leftarrow T \cup \{(u, v)\}$
            UNION(x, y)

Figure 2.5 illustrates the execution of the KRUSKAL algorithm on the graph from Figure 2.1. Initially, every vertex is a tree in the forest. Let the sorted order of the edges be $\langle (d,e), (g,h), (e,f), (d,f), (b,d), (e,g), (f,h), (b,c), (a,b), (c,h), (a,g) \rangle$. Since (d, e) joins two distinct trees in the forest, it is added to the forest, thereby merging the two trees (see Figure 2.5(a)). Next we consider (g, h). Vertex g and vertex h belong to two different trees, thus (g, h) is added to the forest as shown in Figure 2.5(b). In Figure 2.5(d), when (d, f) is processed, both d and f belong to the same tree, therefore we do nothing for this edge. The final minimum spanning tree is shown in Figure 2.5(h).

Sorting the edges in nondecreasing order takes $O(m \log m)$ time. The total running time of determining if the edge joins two distinct trees in the forest is $O(m\alpha(m, n))$ time, where α is the functional inverse of Ackermann's function defined in [85]. Therefore the asymptotic running time of Kruskal's algorithm is $O(m \log m)$, which is the same as $O(m \log n)$ since $\log m = \Theta(\log n)$ by observing that $m = O(n^2)$ and $m = \Omega(n)$.

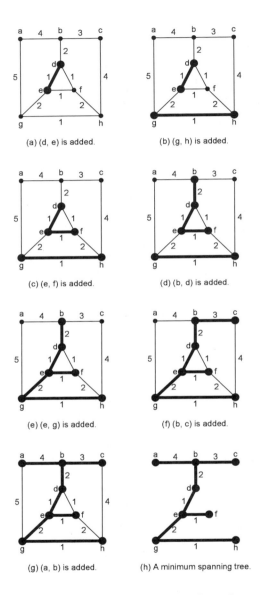

FIGURE 2.5: The execution of the KRUSKAL algorithm on the graph from Figure 2.1.

2.5 Applications

Minimum spanning trees are useful in constructing networks, by describing the way to connect a set of sites using the smallest total amount of wire. Much of the work on minimum spanning trees has been conducted by the communications company.

2.5.1 Cable TV

One example is a cable TV company laying cable to a new neighborhood. If it is constrained to bury the cable only along certain paths, then there would be a graph representing which points are connected by those paths. Some of those paths might be more expensive, because they are longer, or require the cable to be buried deeper. A spanning tree for that graph would be a subset of those paths that has no cycles but still connects to every house. There might be several spanning trees possible. A minimum spanning tree would be one with the lowest total cost.

2.5.2 Circuit design

In the design of electronic circuitry, it is often necessary to wire some pins together in order to make them electrically equivalent. A minimum spanning tree needs the least amount of wire to interconnect a set of points.

2.5.3 Islands connection

Suppose we have a group of islands that we wish to link with bridges so that it is possible to travel from one island to any other in the group. Further suppose that the government wishes to spend the minimum amount on this project. The engineers are able to calculate a cost for a bridge linking each possible pair of islands. The set of bridges that will enable one to travel from any island to any other at the minimum cost to the government is the minimum spanning tree.

2.5.4 Clustering gene expression data

Minimum spanning trees also provide a reasonable way for clustering points in space into natural groups. For example, Ying Xu and his coworkers [103] describe a new framework for representing a set of multi-dimensional gene expression data as a minimum spanning tree. A key property of this representation is that each cluster of the gene expression data corresponds to one subtree of the MST, which rigorously converts a multi-dimensional clustering problem to a tree partitioning problem. They have demonstrated that, although the inter-data relationship is greatly simplified in the MST representation, no essential information is lost for the purpose of clustering. They observe that there are two key advantages in representing a set of multi-dimensional data as an MST. One is that the simple structure of a tree facilitates efficient implementations of rigorous clustering algorithms, which otherwise are highly computationally challenging. The other is that it can overcome many of the problems faced by classical clustering algorithms since an MST-based clustering does not depend on detailed geometric shape of a cluster. A new software tool called EXCAVATOR, which stands for "EXpression data Clustering Analysis and VisualizATiOn Resource," has been developed based on this new framework. The clustering results on the gene expression data (1) from yeast *Saccharomyces cerevisiae*, (2) in response of human fibroblasts to serum, and (3) of *Arabidopsis* in response to chitin elicitation are very promising.

2.5.5 MST-based approximations

In the traveling salesperson problem (TSP), we are given a complete undirected graph G that has weight function w associated with each edge, and we wish to find a tour of G with minimum weight. This problem has been shown to be NP-hard even when the weight function satisfies the triangle inequality, i.e., for all three vertices $x, y, z \in V$, $w(x,z) \leq w(x,y) + w(y,z)$. The triangle inequality arises in many practical situations. It can be shown that the following strategy delivers an approximation algorithm with a ratio bound of 2 for the traveling salesperson problem with triangle inequality. First, find a minimum spanning tree T for the given graph. Then double the MST and construct a tour T'. Finally, add shortcuts so that no vertex is visited more than once, which is done by a preorder tree walk. The resulting tour is of length no more than twice of the optimal. Chapter 7 shows that an MST-based approach also provides a good approximation for the Steiner tree problems.

2.6 Summary

We have briefly discussed three well-known algorithms for solving the minimum spanning tree problem: Borůvka's algorithm, Prim's algorithm, and Kruskal's algorithm. All of them work in a "greedy" fashion.

For years, many improvements have been made for this classical problem. We close this chapter by sketching one possible extension to these three basic algorithms. First apply the contraction step in Borůvka's algorithm for $O(\log \log n)$ time. This takes $O(m \log \log n)$ time since each Borůvka's contraction step takes $O(m)$ time. After these contraction steps, the number of the supervertices of the contracted graph is at most $O(n/2^{\log \log n}) = O(n/\log n)$. Then apply Prim's algorithm to the contracted graph, which runs in time $O(m + (n/\log n)\log n) = O(m + n)$. In total, this hybrid algorithm solves the minimum spanning tree problem in $O(m \log \log n)$ time.

Bibliographic Notes and Further Reading

The history of the *minimum spanning tree* (MST) problem is long and rich. An excellent survey paper by Ronald Graham and Pavol Hell [47] describes the history of the problem up to 1985. The earliest known MST algorithm was proposed by Otakar Borůvka [14], a great Czech mathematician, in 1926. At that time, he was considering an efficient electrical coverage of Bohemia, which occupies the western and middle thirds of today's Czech Republic. In the mid-1950s when the computer age just began, the MST problem was attacked again by several researchers. Among them, Joseph Kruskal [69] and Robert Prim [79] gave two commonly used textbook algorithms. Both of them mentioned Borůvka's paper. In fact, Prim's algorithm was a rediscovery of the algorithm by the prominent number theoretician Vojtěch Jarník [59].

Textbook algorithms run in $O(m \log n)$ time. Andrew Chi-Chih Yao [104], and David Cheriton and Robert Tarjan [20] independently made improvements to $O(m \log \log n)$. By the invention of Fibonacci heaps, Michael Fredman and Robert Tarjan [38] reduced the complexity to $O(m\beta(m,n))$, where $\beta(m,n) = \min\{i|\log^i n \leq m/n\}$. In the worst case, $m = O(n)$ and the running time is $O(m \log^* m)$. The complexity was further lowered to $O(m \log \beta(m,n))$ by Harold N. Gabow, Zvi Galil, Thomas H. Spencer, and Robert Tarjan [40].

On the other hand, David Karger, Philip Klein, and Robert Tarjan [61] gave a randomized linear-time algorithm to find a minimum spanning tree in the restricted random-access model. If the edge costs are integer and the models allow bucketing and bit manipulation, Michael Fredman and Dan Willard [37] gave a deterministic linear-time algorithm.

Given a spanning tree, how fast can we verify that it is minimum? Robert Tarjan [86] gave an almost linear-time algorithm by using path compression. János Komlós [67] showed that a minimum spanning tree can be verified in linear number of comparisons, but with nonlinear overhead to decide which comparisons to make. Brandon Dixon, Monika Rauch, and Robert Tarjan [30] gave the first linear-time verification algorithm. Valerie King [66] proposed a simpler linear-time verification algorithm. All these methods use the fact that a spanning tree is a minimum spanning tree if and only if the weight of each nontree edge (u,v) is at least the weight of the heaviest edge in the path in the tree between u and v.

It remains an open problem whether a linear-time algorithm exists for finding a minimum spanning tree. Bernard Chazelle [18] took a significant step towards a solution and charted out a new line of attack. His algorithm runs in $O(m\alpha(m,n))$ time, where α is the functional inverse of Ackermann's function defined in [85]. The key idea is to compute suboptimal independent sets in a nongreedy fashion, and then progressively improve upon them until an optimal solution is reached. An approximate priority queue, called a *soft heap* [19], is used to construct a suboptimal spanning tree, whose quality is progressively refined until a minimum spanning tree is finally produced.

Seth Pettie and Vijaya Ramachandran [76] established that the algorithmic complexity of the minimum spanning tree problem is equal to its decision-tree complexity. They gave a deterministic, comparison-based MST algorithm that runs in $O(T^*(m,n))$, where $T^*(m,n)$ is the number of edge-weight comparisons needed to determine the MST. Because of the nature of their algorithm, its exact running time is unknown. The source of their algorithm's mysterious running time, and its optimality, is the use of precomputed "MST decision trees" whose exact depth is unknown but nonetheless provably optimal. A trivial lower bound is $\Omega(m)$; and the best upper bound is $O(m\alpha(m,n))$ [18].

Exercises

2-1. Prove that if the weights on the edges of a connected graph are distinct, then there is a unique minimum spanning tree.

2-2. Prove or disprove that if unique, the shortest edge is included in any minimum spanning tree.

2-3. Let e be a minimum-weight edge in a graph G. Show that e belongs to some minimum spanning tree of G.

2-4. Let e be a maximum-weight edge on some cycle of $G = (V, E)$ and $G' = (V, E - \{e\})$. Show that a minimum spanning tree of G' is also a minimum spanning tree of G.

2-5. Devise an algorithm to determine the smallest change in edge cost that causes a change of the minimum spanning tree.

2-6. Given a graph, its minimum spanning tree, and an additional vertex with its associated edges and edge costs, devise an algorithm for rapidly updating the minimum spanning tree.

2-7. Prove or disprove that Borůvka's, Kruskal's, and Prim's algorithms still apply even when the weights may be negative.

2-8. Devise an algorithm to find a *maximum* spanning tree of a given graph. How efficient is your algorithm?

2-9. Devise an algorithm to find a minimum spanning forest, under the restriction that a specified subset of the edges must be included. Analyze its running time.

Chapter 3

Shortest-Paths Trees

3.1 Introduction

Given a road map of the United States on which the distance between each pair of adjacent intersections is marked, how can a motorist determine the shortest route from New York City to San Francisco? The brute-force way is to generate all possible routes from New York City to San Francisco, and select the shortest one among them. This approach apparently generates too many routes that are not worth considering. For example, a route from New York City to Miami to San Francisco is a poor choice. In this chapter we introduce some efficient algorithms for finding all the shortest paths from a given starting location.

Consider a connected, undirected network with one special node, called the source (or root). Associated with each edge is a distance, a nonnegative number. The objective is to find the set of edges connecting all nodes such that the sum of the edge lengths from the source to each node is minimized. We call it a *shortest-paths tree* (SPT) rooted at the source.

In order to minimize the total path lengths, the path from the root to each node must be a shortest path connecting them. Otherwise, we substitute such a path with a shortest path, and get a "lighter" spanning tree whose total path lengths from the root to all nodes are smaller.

Shortest-paths trees are not necessarily unique. Figure 3.1 gives two shortest-paths trees rooted at vertex a for the graph from Figure 2.1. Take a look at the paths from a to e. In Figure 3.1(a), it goes from a to g, and then g to e. In Figure 3.1(b), it goes from a to b, b to d, and then d to e. Both of them are of length 7, which is the length of a shortest path from a to e. Notice that the total edge weight of two shortest-paths trees may not be the same. For example, the total edge weight of the shortest-paths tree in Figure 3.1(a) is 18, whereas that of the shortest-paths tree in Figure 3.1(b) is 17. In Chapter 2, we have shown that the total edge weight of a minimum spanning tree (see Figure 2.2) is 14.

As long as all the edge weights are nonnegative, the shortest-paths tree is well defined. Unfortunately, things get somewhat complicated in the presence of negative edge weights. For an undirected graph, a long path gets "shorter" when we repeatedly add an edge with negative weight to it. In this situation,

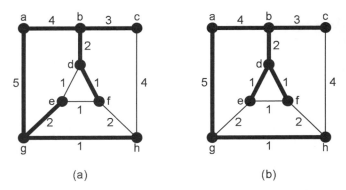

FIGURE 3.1: A shortest-paths tree rooted at vertex a for the graph from Figure 2.1.

a shortest path that contains an edge with negative weight is not well defined since a lesser-weight path can always be found by going back and forth on the negative-weight edge. Consider the graph in Figure 3.2. In this graph, edge (d, e) is of negative weight. Since, for an undirected graph, an edge can be traversed in both directions, a path that repeatedly uses (d, e) will reduce its length. However, for a directed graph, as long as there exists no negative-weight cycle reachable from the source, the shortest-path weights are well defined. Thus when talking about the topics related to shortest paths, we usually focus on solving problems in directed graphs. It should be noted, however, that most such algorithms can be easily adapted for undirected graphs.

In this chapter, the terms "edge" and "arc" are used interchangeably. We discuss two well-known algorithms for constructing a shortest-paths tree: Dijkstra's algorithm and the Bellman-Ford algorithm. Dijkstra's algorithm assumes that all edge weights in the graph are nonnegative, whereas the Bellman-Ford algorithm allows negative-weight edges in the graph. If there is

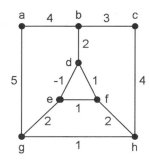

FIGURE 3.2: An undirected graph with a negative-weight edge.

no negative-weight cycle, the Bellman-Ford algorithm produces the shortest paths and their weights. Otherwise, the algorithm detects the negative cycles and indicates that no solution exists.

3.2 Dijkstra's Algorithm

Djikstra's algorithm solves the problem of finding the shortest path from a source to a destination. It turns out that one can find the shortest paths from a given source to all vertices in a graph in the same time; hence, this problem is sometimes called the single-source shortest paths problem. In fact, this algorithm can be used to deliver the set of edges connecting all vertices such that the sum of the edge lengths from the source to each node is minimized.

For each vertex $v \in V$, Dijkstra's algorithm maintains an attribute $\delta[v]$, which is an upper bound on the weight of a shortest path from the source to v. We call $\delta[v]$ a *shortest-path estimate*. Initially, the shortest-path estimates of all vertices other than the source are set to be ∞. Dijkstra's algorithm also maintains a set S of vertices whose final shortest-path weights from the source have not yet been determined. The algorithm repeatedly selects the vertex $u \in S$ with the minimum shortest-path estimate, and re-evaluates the shortest-path estimates of the vertices adjacent to u. The re-evaluation is often referred to as a *relaxation* step. Once a vertex is removed from S, its shortest-path weight from the source is determined and finalized.

Algorithm: DIJKSTRA
Input: A weighted, directed graph $G = (V, E, w)$; a source vertex s.
Output: A shortest-paths spanning tree T rooted at s.
    for each vertex $v \in V$ **do**
        $\delta[v] \leftarrow \infty$
        $\pi[v] \leftarrow$ NIL
    $\delta[s] \leftarrow 0$
    $T \leftarrow \emptyset$
    $S \leftarrow V$
    while $S \neq \emptyset$ **do**
        Choose $u \in S$ with minimum $\delta[u]$
        $S \leftarrow S - \{u\}$
        if $u \neq s$ **then** $T \leftarrow T \cup \{(\pi[u], u)\}$
        for each vertex v adjacent to u **do**
            if $\delta[v] > \delta[u] + w(u, v)$ **then**
                $\delta[v] \leftarrow \delta[u] + w(u, v)$
                $\pi[v] \leftarrow u$

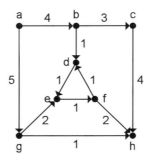

FIGURE 3.3: A weighted, directed graph.

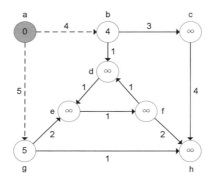

FIGURE 3.4: Vertex a is chosen, and edges (a, b) and (a, g) are relaxed.

Consider the graph in Figure 3.3. The following figures illustrate how Dijkstra's algorithm works in constructing a shortest-paths tree rooted at vertex a.

Initially, all $\delta[\cdot]$ values are ∞, except $\delta[a] = 0$. Set S contains all vertices in the graph. Vertex a, shown as a shaded vertex in Figure 3.4, has the minimum δ value and is chosen as vertex u in the **while** loop. We remove a from S. Then edges (a, b) and (a, g), shown as dotted lines, are relaxed.

Now shaded vertex b in Figure 3.5 has the minimum δ value among all vertices in S and is chosen as vertex u in the **while** loop. We remove b from S. Dark edge (a, b) is added to the shortest-paths tree under construction, and dotted edges (b, c) and (b, d) are relaxed.

Both vertices d and g in Figure 3.6 have the minimum δ values among all vertices in S. Let us choose g as vertex u in the **while** loop. We remove g from S. Dark edge (a, g) is added to the shortest-paths tree, and dotted edges (g, e) and (g, h) are relaxed.

Since vertex d in Figure 3.7 has the minimum δ value among all vertices in S, it is chosen as vertex u in the **while** loop. We remove d from S. Dark edge

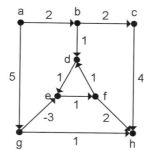

FIGURE 3.13: A weighted, directed graph with a negative-weight edge.

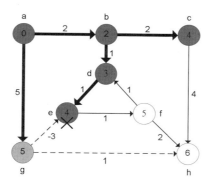

FIGURE 3.14: A failure of Dijkstra's algorithm.

found by using binary heaps in $O(\log n)$ time, runs in $O(m \log n)$ time. By using Fibonacci heaps, we can achieve a running time of $O(m + n \log n)$.

Figure 3.13 gives an example of a directed graph with negative-weight edges for which Dijkstra's algorithm produces an incorrect answer.

As Dijkstra's algorithm proceeds on the graph in Figure 3.13, it will reach the status shown in Figure 3.14. Since edge (g, e) is of negative weight -3, the weight of the path (a, g, e) is $5 - 3 = 2$. It is better than the weight of the finalized shortest-path (a, b, d, e), which is of total weight 4. Furthermore, edge (e, f) has been relaxed. There is no way for Dijkstra's algorithm to re-relax all edges affected by this domino effect.

The graph in Figure 3.15 is a wrong shortest-path tree produced by Dijkstra's algorithm. Notice that (g, e) is of negative weight, and is missing in the shortest-paths tree constructed by Djikstra's algorithm.

A correct shortest-paths tree is given in Figure 3.16.

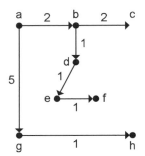

FIGURE 3.15: A wrong shortest-paths tree.

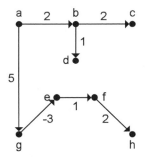

FIGURE 3.16: A correct shortest-paths tree.

3.3 The Bellman-Ford Algorithm

Shortest-paths algorithms typically exploit the property that a shortest path between two vertices contains other shortest paths within it. This optimal substructure property admits both dynamic programming and the greedy method. We have shown that Dijkstra's algorithm might fail in delivering a shortest-paths tree when the graph has some negative-weight edges. The Bellman-Ford algorithm works in a more-general case than Dijkstra's algorithm. It solves the shortest-paths tree problem even when the graph has negative-weight edges. If there is a negative-weight cycle, the algorithm indicates that no solution exists.

Recall that Dijkstra's algorithm only relaxes those edges incident to the chosen vertex with the minimum shortest-path estimate. The failure of Dijkstra's algorithm for graphs with negative-weight edges is due to the fact that it does not calculate the domino effect caused by negative edges. Take a look at Figure 3.14 once again. In that figure, if we push the negative-edge effect forward, a correct shortest-paths tree can then be built. This inspires the design of the Bellman-Ford algorithm which relaxes all edges in each iteration.

Initially, the shortest-path estimates of all vertices other than the source are set to be ∞. Then the algorithm makes $n-1$ passes over all the edges of the graph. In each pass, if $\delta[v] > \delta[u] + w(u,v)$, then we set the value of $\delta[v]$ to be $\delta[u] + w(u,v)$, and modify the predecessor of vertex v. A notable feature of the Bellman-Ford algorithm is that in the kth iteration, the shortest-path estimate for vertex v, i.e., $\delta[v]$, equals the length of the shortest path from the source to v with at most k edges. If, after $n-1$ passes, there exists an edge (u,v) such that $\delta[v] > \delta[u] + w(u,v)$, then a negative cycle has been detected. Otherwise, for all vertices other than the source, we build the shortest-paths tree by adding the edges from their predecessors to them.

Algorithm: BELLMAN-FORD
Input: A weighted, directed graph $G = (V, E, w)$; a source vertex s.
Output: A shortest-paths spanning tree T rooted at s.
    for each vertex $v \in V$ **do**
        $\delta[v] \leftarrow \infty$
        $\pi[v] \leftarrow$ NIL
    $\delta[s] \leftarrow 0$
    for $i \leftarrow 0$ **to** $n-1$ **do**
        for each $(u,v) \in E$ **do**
            if $\delta[v] > \delta[u] + w(u,v)$ **then**
                $\delta[v] \leftarrow \delta[u] + w(u,v)$
                $\pi[v] \leftarrow u$
    for each $(u,v) \in E$ **do**

if $\delta[v] > \delta[u] + w(u,v)$ **then**
   Output "A negative cycle exists."
   Exit
$T \leftarrow \emptyset$
for $v \in V - s$ **do**
   $T \leftarrow T \cup \{(\pi[v], v)\}$

The Bellman-Ford algorithm has the running time of $O(mn)$ since there are $O(n)$ iterations, and each iteration takes $O(m)$ time. The correctness follows from the fact that in the kth iteration, the shortest-path estimate for each vertex equals the length of the shortest path from the source to that vertex with at most k edges. Since a simple path in G contains at most $n-1$ edges, the shortest-path estimates stabilize after $n-1$ iterations unless there exists a negative cycle in the graph.

The following figures show how the execution of the BELLMAN-FORD works on the graph in Figure 3.13. The dark dotted edges record those that do cause some effect in the relaxation. Initially, $\delta[a] = 0$. In Figure 3.17, the relaxation of edge (a,b) changes $\delta[b]$ from ∞ to 2; and the relaxation of edge (a,g) changes $\delta[g]$ from ∞ to 5. Shortest-path estimates $\delta[a]$, $\delta[b]$, and $\delta[g]$ are finalized. Shortest-path estimates $\delta[c]$, $\delta[d]$, $\delta[e]$, $\delta[f]$, and $\delta[h]$ are still ∞ since their corresponding vertices cannot be reached from vertex a by a path with only one edge.

In Figure 3.18, the relaxation of edge (b,c) changes $\delta[c]$ from ∞ to 4; the relaxation of edge (b,d) changes $\delta[d]$ from ∞ to 3; the relaxation of edge (g,e) changes $\delta[e]$ from ∞ to 2; and the relaxation of edge (g,h) changes $\delta[h]$ from ∞ to 6. Up to this stage, shortest-path estimates $\delta[a]$, $\delta[b]$, $\delta[c]$, $\delta[d]$, $\delta[e]$, and $\delta[g]$ are finalized. Shortest-path estimate $\delta[f]$ is still ∞ since vertex f cannot be reached from vertex a by a path with at most two edges, and shortest-path estimate $\delta[h]$ will be modified later.

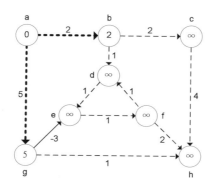

FIGURE 3.17: Shortest-path estimates $\delta[b]$ and $\delta[g]$ are modified.

Shortest-Paths Trees

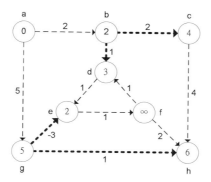

FIGURE 3.18: Shortest-path estimates $\delta[c]$, $\delta[d]$, $\delta[e]$, and $\delta[h]$ are modified.

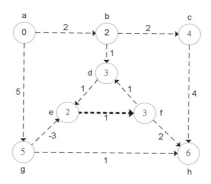

FIGURE 3.19: Shortest-path estimate $\delta[f]$ is modified.

In Figure 3.19, the relaxation of edge (e, f) changes $\delta[f]$ from ∞ to 3.
In Figure 3.20, the relaxation of edge (f, h) changes $\delta[h]$ from 6 to 5.
Figure 3.21 gives the final shortest-paths tree constructed by the Bellman-Ford algorithm.

3.4 Applications

The shortest-paths tree problem comes up in practice and arises as a subproblem in many network optimization algorithms. The shortest path tree is widely used in IP multicast and in some of the application-level multicast routing algorithms.

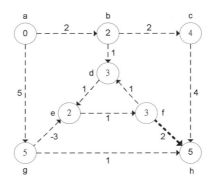

FIGURE 3.20: Shortest-path estimate $\delta[h]$ is modified.

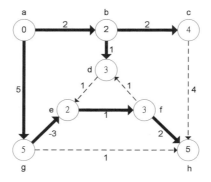

FIGURE 3.21: The shortest-paths tree constructed by the Bellman-Ford algorithm.

3.4.1 Multicast

In the age of multimedia and high-speed networks, multicast is one of the mechanisms by which the power of the Internet can be further utilized in an efficient manner. When more than one receiver is interested in receiving a transmission from a single or a set of senders, multicast is the most efficient and viable mechanism.

Multicast routing refers to the construction of a tree rooted at the source and spanning all destinations. Generally, there are two types of such a tree, Steiner trees (see Chapter 7) and shortest-paths trees. Steiner trees or group-shared trees tend to minimize the total cost of the resulting trees. Shortest-paths trees or source-based trees tend to minimize the cost of each path from source to any destination.

In the following, we briefly describe two well-known routing protocols: the Routing Information Protocol (RIP) and Open Shortest Path First (OSPF).

RIP is a distance-vector protocol that allows routers to exchange information about destinations for computing routes throughout the network. Destinations may be networks or a special destination used to convey a default route. In RIP, the Bellman-Ford algorithms make each router periodically broadcast its routing tables to all its neighbors. Then a router knowing its neighbors' tables can decide to which destination neighbor to forward a packet.

OSPF is a routing protocol developed for Internet Protocol (IP) networks by the Interior Gateway Protocol (IGP) Working Group of the Internet Engineering Task Force (IETF). OSPF was created because in the mid-1980s, RIP was increasingly incapable of serving large, heterogeneous internetworks. Like most link-state algorithms, OSPF uses a graph-theoretic model of network topology to compute shortest paths. Each router periodically broadcasts information about the status of its connections. OSPF floods information about adjacencies to all routers in the network where each router locally computes the shortest paths by running Dijkstra's algorithm.

3.4.2 SPT-based approximations

Besides their applications to network routing problems, the shortest-paths tree algorithms could also serve as good approximations for some NP-hard problems. For example, in Chapter 4 we will show that a shortest-paths tree rooted at some vertex is a 2-approximation of the minimum routing cost spanning tree (MRCT) problem, which is known to be NP-hard. In fact, several SPT-based approximations will be studied in the next chapters.

3.5 Summary

We have introduced two most basic algorithms for constructing a shortest-paths tree for a given directed or undirected weighted graph. Both of them use the technique of relaxation, progressively decreasing a shortest-path estimate $\delta[v]$ for each vertex v. The relaxation causes the shortest-path estimates to descend monotonically toward the actual shortest-path weights. Dijkstra's algorithm relaxes each edge exactly once (twice in the case of undirected graphs) if all the edge weights are positive. On the other hand, the Bellman-Ford algorithm relaxes each edge $n-1$ times, so that the effect of a negative edge can be propagated properly. If the shortest-path estimates do not stabilize after $n-1$ passes, then there must exist a negative cycle in the graph, and the algorithm indicates that no solution exists.

Bibliographic Notes and Further Reading

The shortest-paths tree problem is one of the most classical network flow optimization problems. An equivalent problem is to find a shortest path from a given source vertex $s \in V$ to every vertex $v \in V$. Algorithms for this problem have been studied for a long time. In fact, since the end of the 1950s, thousands of scientific works have been published. A good description of the classical algorithms and their implementations can be found in [21, 42].

Dijkstra's original algorithm, by Edsger W. Dijkstra [28], did not mention the usage of a priority queue. A discussion of using different priority queue techniques can be found in [21, 25].

For the shortest path problem with nonnegative arc lengths, the Fibonacci heap data structure [38] yields an $O(m+n \log n)$ implementation of Dijkstra's algorithm in the pointer model of computation. Let U denote the biggest arc length, and C be the ratio between U and the smallest nonzero arc length. In a RAM model with word operations, the fastest known algorithms achieve the following bounds: $O(m + n(\sqrt{\log n})$ [81], $O(m + n(\log C \log \log C)^{1/3})$ [46, 82], $O(m \log \log U)$ [50], and $O(m \log \log n)$ [88]. Ulrich Meyer [74] shows the problem can be solved in linear average time if input arc lengths are independent and uniformly distributed. Andrew V. Goldberg [46] shows that a simple modification of the algorithm of [27] yields an algorithm with linear average running time on the uniform arc length distribution. For undirected graphs, Mikkel Thorup [87] gave a linear-time algorithm in a word RAM model.

The Bellman-Ford algorithm is based on separate algorithms by Richard Bellman [10], and Lestor Ford, Jr. and D.R. Fulkerson [36]. Though the

Bellman-Ford algorithm is simple and has a high running time, to date there is no algorithm which significantly improves its asymptotic complexity.

Exercises

3-1. Prove or disprove that, if unique, the shortest edge is included in any shortest-paths tree.

3-2. Show that Kruskal's, Prim's, and Dijkstra's algorithms still apply even when the problem statement requires the inclusion of specific edges.

3-3. Apply the Bellman-Ford algorithm to Figure 3.22, and show how it detects the negative cycle in the graph.

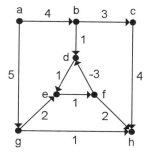

FIGURE 3.22: A directed graph with a negative-weight cycle.

3-4. Given a directed or undirected graph, devise an $O(m+n)$ time algorithm that detects whether there exists a cycle in the graph.

3-5. Devise an algorithm that determines the number of different shortest paths from a given source to a given destination.

3-6. Give an example of a graph G and a vertex v, such that the minimum spanning tree of G is exactly the shortest-paths tree rooted at v.

3-7. Give an example of a graph G and a vertex v, such that the minimum spanning tree of G is very different from the shortest-paths tree rooted at v.

3-8. Let T be the shortest-paths tree of a graph G rooted at v. Suppose now that all the weights in G are increased by a constant number. Is T still the shortest-paths tree?

3-9. Devise an algorithm that finds a shortest path from u to v for given vertices u and v. If we solve the shortest-paths tree problem with root u, we solve this problem, too. Can you find an asymptotically faster algorithm than the best shortest-paths tree algorithm in the worst case?

3-10. Compare Dijkstra's algorithm with Prim's algorithm.

3-11. If the input graph is known to be a directed acyclic graph, can you improve the performance of the Bellman-Ford algorithm?

Chapter 4

Minimum Routing Cost Spanning Trees

4.1 Introduction

Consider the following problem in network design: given an undirected graph with nonnegative delays on the edges, the goal is to find a spanning tree such that the average delay of communicating between any pair using the tree is minimized. The delay between a pair of vertices is the sum of the delays of the edges in the path between them in the tree. Minimizing the average delay is equivalent to minimizing the total delay between all pairs of vertices using the tree.

In general, when the cost on an edge represents a price for routing messages between its endpoints (such as the delay), the *routing cost* for a pair of vertices in a given spanning tree is defined as the sum of the costs of the edges in the unique tree path between them. The routing cost of the tree itself is the sum over all pairs of vertices of the routing cost for the pair in this tree, i.e., $C(T) = \sum_{u,v} d_T(u,v)$, where $d_T(u,v)$ is the distance between u and v on T. For an undirected graph, the *minimum routing cost spanning tree* (MRCT) is the one with minimum routing cost among all possible spanning trees.

Unless specified explicitly in this and the remaining chapters, a graph G is assumed to be simple and undirected, and the edge weights are nonnegative. Finding an MRCT in a general edge-weighted undirected graph is known to be NP-hard. In this chapter, we shall focus on the approximation algorithms. Before going into the details, we introduce a term, *routing load*, which provides us an alternative formula to compute the routing cost of a tree.

DEFINITION 4.1 *Let T be a tree and $e \in E(T)$. Assume X and Y are the two subgraphs that result by removing e from T. The routing load on edge e is defined by $l(T,e) = 2|V(X)| \times |V(Y)|$.*

For any edge $e \in E(T)$, let x and y ($x \leq y$) be the numbers of vertices in the two subtrees that result by removing e. The routing load on e is $2xy = 2x(|V(T)| - x)$. Note that $x \leq n/2$, and the routing load increases as x increases. The following property can be easily shown by the definition.

FACT 4.1
For any edge $e \in E(T)$, if the numbers of vertices in both sides of e are at least $\delta|V(T)|$, the routing load on e is at least $2\delta(1-\delta)|V(T)|^2$. Furthermore, for any edge of a tree T, the routing load is upper bounded by $\frac{|V(T)|^2}{2}$.

For a graph G and $u, v \in V(G)$, we use $SP_G(u, v)$ to denote a shortest path between u and v in G. In the case where G is a tree, $SP_G(u, v)$ denotes the unique simple path between the two vertices. By defining the routing load, we have the following formula.

LEMMA 4.1
For a tree T with edge length w, $C(T) = \sum_{e \in E(T)} l(T, e) w(e)$. In addition, $C(T)$ can be computed in $O(n)$ time.

PROOF Let $SP_T(u, v)$ denote the simple path between vertices u and v on a tree T.

$$C(T) = \sum_{u,v \in V(T)} d_T(u,v)$$
$$= \sum_{u,v \in V(T)} \left(\sum_{e \in SP_T(u,v)} w(e) \right)$$
$$= \sum_{e \in E(T)} \left(\sum_{u \in V(T)} |\{v | e \in SP_T(u,v)\}| \right) w(e)$$
$$= \sum_{e \in E(T)} l(T,e) w(e).$$

To compute $C(T)$, it is sufficient to find the routing load on each edge. This can be done in $O(n)$ time by rooting T at any node and traversing T in a postorder sequence. □

Example 4.1
Consider the tree T in Figure 4.1. The distances between vertices are as follows:

$$\begin{array}{lll}
d_T(v_1, v_2) = 10 & d_T(v_1, v_3) = 5 & d_T(v_1, v_4) = 13 \\
d_T(v_1, v_5) = 11 & d_T(v_2, v_3) = 15 & d_T(v_2, v_4) = 3 \\
d_T(v_2, v_5) = 1 & d_T(v_3, v_4) = 18 & d_T(v_3, v_5) = 16 \\
d_T(v_4, v_5) = 4 & &
\end{array}$$

The routing cost of T is two times the sum of the above distances since, for v_i and v_j, both $d_T(v_i, v_j)$ and $d_T(v_j, v_i)$ are counted in the cost. We have $C(T) = 192$.

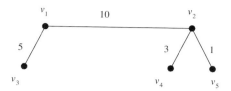

FIGURE 4.1: An example illustrating the routing loads.

On the other hand, the routing load of edge (v_1, v_2) is calculated by

$$l(T, (v_1, v_2)) = 2 \times 2 \times 3 = 12$$

since there are two and three vertices in the two sides of the edge, respectively. Similarly, the routing loads of the other edges are as follows:

$$l(T, (v_1, v_3)) = 8 \quad l(T, (v_2, v_4)) = 8 \quad l(T, (v_2, v_5)) = 8.$$

Therefore, by Lemma 4.1, we have

$$C(T) = 12 \times 10 + 8 \times 5 + 8 \times 3 + 8 \times 1 = 192.$$

□

At the first sight of the definition $C(T) = \sum_{u,v} d_T(u,v)$, one may think that the weights of the tree edges are the most important to the routing cost of a tree. On the other hand, by Lemma 4.1, one can see that the routing loads also play important roles. The routing loads are determined by the topology of the tree. Therefore the topology is crucial for constructing a tree of small routing cost. The next example illustrates the impact of the topology by considering two extreme cases.

Example 4.2

Let T_1 be a star (a tree with only one internal node) in which each edge has weight 5, and T_2 be a path in which each edge has weight 1 (Figure 4.2). Suppose that both the two trees are spanning trees of a graph with n vertices. In the aspect of total edge weight, we can see that T_2 is better than T_1. Let's compute their routing costs. For T_1, the routing load of each edge is $2(n-1)$ since each edge is incident with a leaf. Therefore, by Lemma 4.1, $C(T_1) = 10(n-1)^2$.

Let $T_2 = (v_1, v_2, \ldots, v_n)$. Removing an edge (v_i, v_{i+1}) will result in two components of i and $n-i$ vertices. Therefore the routing loads are

$$2(n-1), 2 \times 2 \times (n-2), \ldots, 2 \times i \times (n-i), \ldots, 2(n-1).$$

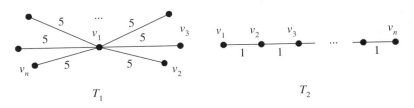

FIGURE 4.2: Two extreme trees illustrating the impact of the topology.

By Lemma 4.1,

$$C(T_2) = \sum_{1 \leq i \leq n-1} 2i(n-i) = n^2(n-1) - \frac{n(n-1)(2n-1)}{3}$$
$$= \frac{n(n-1)(n+1)}{3}.$$

So T_2 is much more costly than T_1 when n is large. □

In the following sections, we shall discuss the complexity and several approximation algorithms for the MRCT problem. In this chapter, by mrct(G), we denote a minimum routing cost spanning tree of a graph G. When there is no ambiguity, we assume that $G = (V, E, w)$ is the given underlying graph, which is simple, connected, and undirected. We shall also use $\widehat{T}$ and mrct(G) interchangeably.

4.2 Approximating by a Shortest-Paths Tree

4.2.1 A simple analysis

Since all edges have nonnegative length, the distance between two vertices is not decreased by removing edges from a graph. Obviously $d_T(u,v) \geq d_G(u,v)$ for any spanning tree T of G. We can obtain a trivial lower bound of the routing cost.

FACT 4.2
$C(\widehat{T}) \geq \sum_{u,v \in V} d_G(u,v).$

Let r be the *median* of graph $G = (V, E, w)$, i.e., the vertex with minimum total distance to all vertices. In other words, r minimizing the function $f(v) =$

$\sum_{u \in V} d_G(v, u)$. We can show that a shortest-paths tree rooted at r is a 2-approximation of an MRCT.

THEOREM 4.1
A shortest-paths tree rooted at the median of a graph is a 2-approximation of an MRCT of the graph.

PROOF   Let r be the median of graph $G = (V, E, w)$ and Y be any shortest-paths tree rooted at r. Note that the triangle inequality holds for the distances between vertices in any graph without edges of negative weights. By the triangle inequality, we have $d_Y(u, v) \leq d_Y(u, r) + d_Y(v, r)$ for any vertices u and v. Summing up over all pairs of vertices, we obtain

$$C(Y) = \sum_u \sum_v d_Y(u, v)$$
$$\leq n \sum_u d_Y(u, r) + n \sum_v d_Y(v, r)$$
$$= 2n \sum_v d_Y(v, r).$$

Since r is the median, $\sum_v d_G(r, v) \leq \sum_v d_G(u, v)$ for any vertex u, it follows

$$\sum_v d_G(r, v) \leq \frac{1}{n} \sum_{u,v} d_G(u, v).$$

Recall that, in a shortest-paths tree, the path from the root to any vertex is a shortest path on the original graph. We have $d_Y(r, v) = d_G(r, v)$ for each vertex v, and consequently

$$C(Y) \leq 2n \sum_v d_G(r, v) \leq 2 \sum_{u,v} d_G(u, v).$$

By Fact 4.2, Y is a 2-approximation of an MRCT. □

The median of a graph can be found easily once the distances of all pairs of vertices are known. By Theorem 4.1, we can have a 2-approximation algorithm and the time complexity is dominated by that of finding all-pairs shortest path lengths of the input graph.

COROLLARY 4.1
An MRCT of a graph can be approximated with approximation ratio 2 in $O(n^2 \log n + mn)$ time.

The approximation ratio in Theorem 4.1 is tight in the sense that there exists an extreme case for the inequality asymptotically. Consider a complete

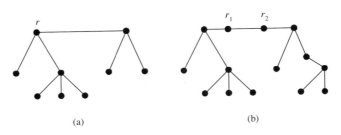

FIGURE 4.3: A centroid of a tree.

graph with unit length on each edge. Any vertex is a median of the graph and the shortest-paths tree rooted at a median is just a star. The routing cost of the star can be calculated by

$$2(n-1)(n-2) + 2(n-1) = 2(n-1)^2.$$

Since the total distance on the graph is $n(n-1)$, the ratio is $2 - (2/n)$.

However, the analysis of the extreme case does not imply that the approximation ratio cannot be improved. It only means that we cannot get a more precise analysis by comparing with the trivial lower bound. It can be easily verified that, for the above example, the star is indeed an optimal solution. The problem is that we should compare the approximation solution with the optimal, but not with the trivial lower bound. Now we introduce another proof of the approximation ratio of the shortest-paths tree. The analysis technique we used is called *solution decomposition*, which is widely used in algorithm design, especially for approximation algorithms.

4.2.2 Solution decomposition

To design an approximation algorithm for an optimization problem, we first suppose that X is an optimal solution. Then we decompose X and construct another feasible solution Y. To our aim, Y is designed to be a good approximation of X and belongs to some restricted subset of feasible solutions, of which the best solution can be found efficiently. The algorithm is designed to find an optimal solution of the restricted problem, and the approximation ratio is ensured by that of Y. It should be noted that Y plays a role only in the analysis of approximation ratio, but not in the designed algorithm. In the following, we show how to design a 2-approximation algorithm by this method.

For any tree, we can always cut it at a node r such that each branch contains at most half of the nodes. Such a node is usually called a *centroid* of the tree in the literature. For example, in Figure 4.3(a), vertex r is a centroid of the tree. The tree has nine vertices. Removing r from the tree will result in three subtrees, and each subtree has no more than four vertices. It can be easily verified that r is the unique centroid of the tree in (a). However, for the tree in (b), both r_1 and r_2 are centroids of the tree.

Suppose that r is the centroid of the MRCT $\widehat{T}$. If we construct a shortest-paths tree Y rooted at the centroid r, the routing cost will be at most twice that of $\widehat{T}$. This can be easily shown as follows. First, if u and v are two nodes not in a same branch, $d_{\widehat{T}}(u,v) = d_{\widehat{T}}(u,r) + d_{\widehat{T}}(v,r)$. Consider the total distance of all pairs of nodes on $\widehat{T}$. For any node v, since each branch contains no more than half of the nodes, the term $d_{\widehat{T}}(v,r)$ will be counted in the total distance at least n times, $n/2$ times for v to others, and $n/2$ times for others to v. Hence, we have $C(\widehat{T}) \geq n \sum_v d_{\widehat{T}}(v,r)$. Since, as in the proof of Theorem 4.1, $C(Y) \leq 2n \sum_v d_G(v,r)$, it follows that $C(Y) \leq 2C(\widehat{T})$. We have decomposed the optimal solution $\widehat{T}$ and construct a 2-approximation Y. Of course, there is no way to know what Y is since the optimal $\widehat{T}$ is unknown. But we have the next result.

LEMMA 4.2
There exists a vertex such that any shortest-paths tree rooted at the vertex is a 2-approximation of the MRCT.

By Lemma 4.2, we can design a 2-approximation algorithm which constructs a shortest-paths tree rooted at each vertex and chooses the best of them. Since there are only n vertices and a shortest-paths tree can be constructed in $O(n \log n + m)$ time, the algorithm runs in polynomial time. By Lemma 4.1, the routing cost of a tree can be computed in $O(n)$ time. Consequently the algorithm has time complexity $O(n^2 \log n + mn)$, the same result as in Corollary 4.1.

In the remaining sections of this chapter, we generalize the above method to improve the approximation ratio. The main idea is the same; however, the proof is much more involved.

4.3 Approximating by a General Star
4.3.1 Separators and general stars

A key point to the 2-approximation in the last section is the existence of the centroid, which separates a tree into sufficiently small components. To generalize the idea, we define the separator of a tree in Definition 4.2.

DEFINITION 4.2 *Let T be a spanning tree of G and S be a connected*

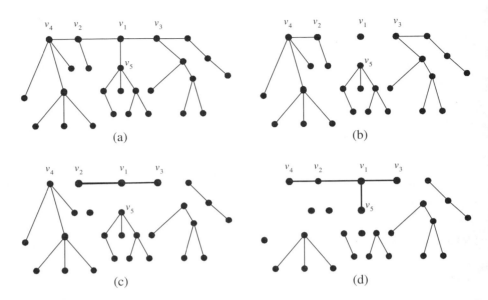

FIGURE 4.4: An example of a minimal separator of a tree.

subgraph of T. A branch of S is a connected component of the subgraph that results by removing S from T.

DEFINITION 4.3 *Let $\delta \leq 1/2$. A connected subgraph S is a δ-separator of T if $|B| \leq \delta |V(T)|$ for every branch B of S.*

A δ-separator S is *minimal* if any proper subgraph of S is not a δ-separator of T.

Example 4.3
The tree in Figure 4.4(a) has 26 vertices in which v_1 is a centroid. The vertex v_1 is a minimal 1/2-separator. As shown in (b), each branch contains no more than 13 vertices. But v_1, or even the edge (v_1, v_2), is not a 1/3-separator because there exists a subtree whose number of vertices is nine, which is greater than 26/3. The path between v_2 and v_3 is a minimal 1/3-separator (Frame (c)), and the subgraph that consists of v_1, v_2, v_3, v_4, and v_5 is a minimal 1/4-separator (Frame (d)). □

The δ-separator can be thought of as a generalization of the centroid of a tree. Obviously, a centroid is a 1/2-separator which contains only one node.

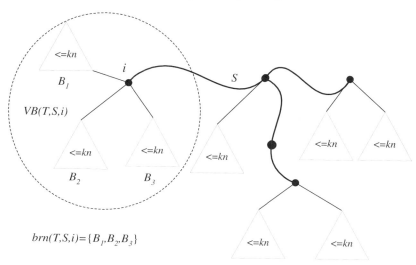

FIGURE 4.5: A δ-separator and branches of a tree. The bold line is the separator S and each triangle is a branch of S.

Intuitively, a separator is like a routing center of the tree. Starting from any node, there are sufficiently many nodes which can only be reached after reaching the separator. For two vertices i and j in different components separated by S, the path between them can be divided into three subpaths: from i to S, a path in S, and from S to j. Since each component contains no more than δn vertices, the distance $d_T(i, S)$ will be counted at least $2(1-\delta)n$ times as we compute the routing cost of T. For each edge e in S, since there are at least δn vertices on either side of the edge, by Fact 4.1, the routing load on e is at least $2\delta(1-\delta)n^2$. Some notations are given below and illustrated in Figure 4.5.

DEFINITION 4.4 *Let T be a spanning tree of G and S be a connected subgraph of T. Let u be a vertex in S. The set of branches of S connected to u by an edge of T is denoted by $brn(T, S, u)$, while $brn(T, S)$ is for the set of all branches of S. The set of vertices in the branches connected to u is denoted by $VB(T, S, u) = \{u\} \cup \{v | v \in B \in brn(T, S, u)\}$.*

The next fact directly follows the definitions.

FACT 4.3
Let S be a minimal δ-separator of T. If v is a leaf of S, then $|VB(T, S, v)| > \delta |V(T)|$.

A star is a tree with only one internal vertex (center). We define a *general star* as follows.

DEFINITION 4.5   *Let R be a tree contained in the underlying graph G. A spanning tree T is a general star with core R if each vertex is connected to R by a shortest path.*

For an extreme example, a shortest-paths tree is a general star whose core contains only one vertex. By $star(R)$, we denote the set of all general stars with core R. The intuition of using general stars to approximate an MRCT is described as follows: Assume S is a δ-separator of an optimal tree T. The separator breaks the tree into sufficiently small components (branches). The routing cost of T is the sum of the distances of the $n(n-1)$ pairs of vertices. If we divide the routing cost into two terms, the total distance of vertices in different branches and the total distance of vertices in a same branch, then the inter-branch distance is the larger fraction of the total routing cost. Furthermore, the fraction gets larger and larger when a smaller δ is chosen. If we construct a general star with core S, the routing cost will be very close to the optimal.

Given a core, to construct a general star is just to find a shortest-paths forest, which can be done in $O(n \log n + m)$ time. However, it can be done more efficiently if the all-pairs shortest paths are given.

LEMMA 4.3
Let G be a graph, and let S be a tree contained in G. A spanning tree $T \in star(S)$ can be found in $O(n)$ time if a shortest path $SP_G(v, S)$ is given for every $v \in V(G)$.

PROOF   A constructive proof is given below. Starting from $T = S$, we show a procedure which inserts all other vertices into T one by one. At each iteration, the following equality is kept:

$$d_T(v, S) = d_G(v, S) \qquad \forall v \in V(T). \tag{4.1}$$

It is easy to see that (4.1) is true initially. Consider the step of inserting a vertex. Let $SP_G(v, S) = (v = v_1, v_2, ..., v_k \in S)$ be a shortest path from v to S, and let v_j be the first vertex which is already in T. Set $T \leftarrow T \cup (v_1, v_2, ..., v_j)$. Since $(v_1, v_2, ..., v_k)$ is a shortest path from v to S, $(v_a, v_{a+1}, ..., v_j)$ is also a shortest path from v_a to v_j for any $a = 1, \ldots j$, and (4.1) is true. It is easy to see that the time complexity is $O(n)$, if a shortest path from v to S is given for every $v \in V$. □

Let S be a connected subgraph of a spanning tree T. The path between two vertices v and u in different branches can be divided into three subpaths:

the path from v to S, the path contained in S, and the path from u to S. For convenience, we define $d_T^S(u,v) = w(SP_T(u,v) \cap S)$. Obviously

$$d_T(u,v) \leq d_T(v,S) + d_T^S(u,v) + d_T(u,S), \tag{4.2}$$

and the equality holds if v and u are in different branches. Summing up (4.2) for all pairs of vertices, we have

$$C(T) \leq 2n \sum_{v \in V} d_T(v,S) + \sum_{u,v \in V} d_T^S(u,v).$$

By the definition of routing load,

$$\sum_{u,v \in V} d_T^S(u,v) = \sum_{e \in E(S)} l(T,e) w(e).$$

Suppose that T is a general star with core S. We can establish an upper bound of the routing cost by observing that $d_T(v,S) = d_G(v,S)$ for any vertex v and $l(T,e) \leq \frac{n^2}{2}$ for any edge e (Fact 4.1).

LEMMA 4.4
Let G be a graph and S be a tree contained in G. If $T \in \text{star}(S)$, $C(T) \leq 2n \sum_{v \in V(G)} d_G(v,S) + (n^2/2) w(S)$.

Now we establish a lower bound of the minimum routing cost. Let S be a minimal δ-separator of a spanning tree T and $\mathcal{X}$ denote the set of the ordered pairs of the vertices not in a same branch of S. For any vertex pair $(u,v) \in \mathcal{X}$,

$$d_T(u,v) = d_T(u,S) + d_T^S(u,v) + d_T(v,S). \tag{4.3}$$

Summing up (4.3) for all pairs in $\mathcal{X}$, we have a lower bound of $C(T)$.

$$C(T) \geq \sum_{(u,v) \in \mathcal{X}} d_T(u,v)$$
$$= \sum_{(u,v) \in \mathcal{X}} (d_T(u,S) + d_T(v,S)) + \sum_{(u,v) \in \mathcal{X}} d_T^S(u,v). \tag{4.4}$$

Since S is a δ-separator, there are at least $(1-\delta)n$ vertices not in the same branch of any vertex v, and we have

$$\sum_{(u,v) \in \mathcal{X}} (d_T(u,S) + d_T(v,S)) \geq 2(1-\delta)n \sum_{v \in V} d_T(v,S). \tag{4.5}$$

Since $d_T^S(u,v) = 0$ if v and u are in the same branch,

$$\sum_{(u,v) \in \mathcal{X}} d_T^S(u,v) = \sum_v \sum_u d_T^S(u,v).$$

By definition, this is the total routing cost on the edges of S. Rewriting this in terms of routing loads, we have

$$\sum_v \sum_u d_T^S(u,v) = \sum_{e \in E(S)} l(T,e)w(e). \qquad (4.6)$$

Substituting (4.5) and (4.6) in (4.4), we have

$$C(T) \geq 2(1-\delta)n \sum_{v \in V} d_T(v, S) + \sum_{e \in E(S)} l(T,e)w(e). \qquad (4.7)$$

Since S is a minimal δ-separator, for any edge of S there are at least δn vertices on either side of the edge. Therefore, $l(T, e) \geq 2\delta(1-\delta)n^2$ for any $e \in E(S)$. Consequently,

$$\sum_{e \in E(S)} l(T,e) w(e) \geq 2\delta(1-\delta)n^2 \sum_{e \in E(S)} w(e) = 2\delta(1-\delta)n^2 w(S). \qquad (4.8)$$

Combining (4.7) and (4.8), we obtain

$$C(T) \geq 2(1-\delta)n \sum_{v \in V} d_T(v, S) + 2\delta(1-\delta)n^2 w(S). \qquad (4.9)$$

Particularly, for the MRCT $\widehat{T}$ we have the next lemma.

LEMMA 4.5
If S is a minimal δ-separator of $\widehat{T}$, then

$$C(\widehat{T}) \geq 2(1-\delta)n \sum_{v \in V} d_{\widehat{T}}(v, S) + 2\delta(1-\delta)n^2 w(S).$$

4.3.2 A 15/8-approximation algorithm

In the last section, a 1/2-separator is used to derive a 2-approximation algorithm. The idea is now generalized to show that a better approximation ratio can be obtained by using a 1/3-separator. The following lemma shows the existence of a 1/3-separator. Note that a path may contain only one vertex.

LEMMA 4.6
For any tree T, there is a path $P \subset T$, such that P is a 1/3-separator of T.

PROOF Let n be the number of vertices of T and r be a centroid of T. There are at most 2 branches of r, in which the number of vertices exceed

$n/3$. If there is no such branch, then r is itself a 1/3-separator. Let A be a branch of r with $|V(A)| > n/3$. Since A itself is a tree with no more than $n/2$ vertices, a centroid r_a of A is a 1/2-separator of A, and each branch of r_a contains no more than $n/4$ vertices of A. If there is another branch B of r such that $|V(B)| > n/3$, a centroid r_b of B can be found such that each branch of r_b contains no more than $n/4$ vertices of B. Consider the path $P = SP_T(r_a, r) \cup SP_T(r, r_b)$. Since each branch of P contains no more than $n/3$ vertices, P is a 1/3-separator of T. Note that if B does not exist, then $SP_T(r_a, r)$ is a 1/3-separator. □

In the following paragraphs, a *path separator* of a tree T is a path and meanwhile a minimal 1/3-separator of T. Substituting $\delta = 1/3$ in Lemma 4.5, we obtain a lower bound of the minimum routing cost.

COROLLARY 4.2
If P is a path separator of $\widehat{T}$, then

$$C(\widehat{T}) \geq \frac{4n}{3} \sum_{v \in V} d_{\widehat{T}}(v, P) + \frac{4n^2}{9} w(P).$$

LEMMA 4.7
There exist $r_1, r_2 \in V$ such that if $R = SP_G(r_1, r_2)$ and $T \in \text{star}(R)$, $C(T) \leq (15/8) C(\widehat{T})$.

PROOF Let P be a path separator of $\widehat{T}$ with endpoints r_1 and r_2. Since T is a general star with core R, by Lemma 4.4,

$$C(T) \leq 2n \sum_{v \in V(G)} d_G(v, R) + \frac{n^2}{2} w(R). \qquad (4.10)$$

Let $S = VB(\widehat{T}, P, r_1) \cup VB(\widehat{T}, P, r_2)$ denote the set of vertices in the branches incident to the two endpoints of P. For any $v \in S$,

$$d_G(v, R) \leq \min\{d_G(v, r_1), d_G(v, r_2)\}$$
$$\leq d_{\widehat{T}}(v, P).$$

For $v \notin S$,

$$d_G(v, R) \leq \min\{d_G(v, r_1), d_G(v, r_2)\}$$
$$\leq (d_G(v, r_1) + d_G(v, r_2))/2$$
$$\leq d_{\widehat{T}}(v, P) + w(P)/2.$$

Then, by Fact 4.3, $|S| \geq \frac{2n}{3}$, and therefore

$$\sum_{v \in V} d_G(v, R) \leq \sum_{v \in V} d_{\widehat{T}}(v, P) + (n/6)w(P). \tag{4.11}$$

Substituting this in (4.10) and recalling that $w(R) \leq w(P)$ since R is a shortest path between r_1 and r_2, we have

$$C(T) \leq 2n \sum_{v \in V} d_{\widehat{T}}(v, P) + (5n^2/6)w(P). \tag{4.12}$$

Comparing with the lower bound in Corollary 4.2, we obtain

$$C(T) \leq \max\{3/2, 15/8\} C(\widehat{T}) = (15/8) C(\widehat{T}).$$

□

By Lemma 4.7 we can have a 15/8-approximation algorithm for the MRCT problem. For every r_1 and r_2 in V, we construct a shortest path $R = SP_G(r_1, r_2)$ and a general star $T \in \text{star}(R)$ including the degenerated cases $r_1 = r_2$. The one with the minimum routing cost must be a 15/8-approximation of the MRCT. All-pairs shortest paths can be found in $O(n^3)$ time. A direct method takes $O(n \log n + m)$ time for each pair r_1 and r_2, and therefore $O(n^3 \log n + n^2 m)$ time in total. In the next lemma, it is shown that this can be done in $O(n^3)$.

LEMMA 4.8
Let $G = (V, E, w)$. There is an algorithm which constructs a general star $T \in \text{star}(SP_G(r_1, r_2))$ for every vertex pair r_1 and r_2 in $O(n^3)$ time.

PROOF For any $r \in V$, if a general star $T \in \text{star}(SP_G(r, v))$ for each $v \in V$ can be constructed with total time complexity $O(n^2)$, then all the stars can be constructed in $O(n^3)$ time by applying the algorithm n times for each $r \in V$. By Lemma 4.3, a star $T \in \text{star}(SP_G(r, v))$ can be constructed in $O(n)$ time if, for every $u \in V$, a shortest path from u to $SP_G(r, v)$ is given. Define $A(u, v) = d_G(u, SP_G(r, v))$ and $B(u, v)$ to be the vertex $k \in SP_G(r, v)$ such that $SP_G(u, k) = SP_G(u, SP_G(r, v))$. Since the all-pairs shortest paths can be constructed in $O(n^2 \log n + mn)$ time at the preprocessing stage, we need to compute $A(u, v)$, as well as $B(u, v)$, in $O(n^2)$ time for all $u, v \in V$.

First, construct a shortest-paths tree S rooted at r. Let $parent(v)$ denote the parent of v in S. It is not hard to see that

$$A(u, v) = \min\{A(parent(v), u), d_G(u, v)\}$$

for $u, v \in V - \{r\}$, and $A(r, u) = d_G(r, u)$. Therefore by a top-down traversal of S, we can compute $A(u, v)$ and $B(u, v)$ for all $u, v \in V$ in $O(n^2)$ time. □

The next theorem can be derived directly from Lemmas 4.7 and 4.8.

THEOREM 4.2
There is a 15/8-approximation algorithm for the MRCT problem with time complexity $O(n^3)$.

4.3.3 A 3/2-approximation algorithm

Let P be a path separator of an optimal tree. By Lemma 4.4, if $X \in \text{star}(P)$, then
$$C(X) \leq 2n \sum_{v \in V} d_G(v, P) + (n^2/2) w(P).$$

Since $d_G(v, P) \leq d_{\widehat{T}}(v, P)$ for any v, it can be shown that X is a 3/2-approximation solution by Lemma 4.2. However, it costs exponential time to try all possible paths. In the following we show that a 3/2-approximation solution can be found if, in addition to the two endpoints of P, we know a centroid of an optimal tree.

Let $P = (p_1, p_2, ..., p_k)$ be a path separator of $\widehat{T}$, $V_i = VB(T, P, p_i)$, and $n_i = |V_i|$ for $1 \leq i \leq k$. It is easy to see that a centroid must be in $V(P)$. Let p_q be a centroid of $\widehat{T}$. Construct $R = SP_G(p_1, p_q) \cup SP_G(p_q, p_k)$. We assume that R has no cycle. Otherwise, we arbitrarily remove edges to break the cycles. Obviously $w(R) \leq w(P)$. Let $T \in \text{star}(R)$. The next lemma shows the approximation ratio.

LEMMA 4.9
$C(T) \leq (3/2) C(\widehat{T})$.

PROOF First, for any $v \in V_1 \cup V_q \cup V_k$,
$$d_G(v, R) \leq \min\{d_G(v, p_1), d_G(v, p_q), d_G(v, p_k)\}$$
$$\leq d_{\widehat{T}}(v, P).$$

For $v \in \bigcup_{1 < i < q} V_i$,
$$d_G(v, R) \leq \min\{d_G(v, p_1), d_G(v, p_q)\}$$
$$\leq (d_G(v, p_1) + d_G(v, p_q))/2$$
$$\leq d_{\widehat{T}}(v, P) + d_{\widehat{T}}(p_1, p_q)/2.$$

Similarly, for $v \in \bigcup_{q < i < k} V_i$,
$$d_G(v, S) \leq d_{\widehat{T}}(v, P) + d_{\widehat{T}}(p_q, p_k)/2.$$

By Fact 4.3 and the property of a centroid, we have $\sum_{1 < i < q} n_i \leq n/6$ and $\sum_{q < i < k} n_i \leq n/6$. Thus,

$$\sum_{v \in V} d_G(v, R) \leq \sum_{v \in V} d_{\widehat{T}}(v, P) + (n/12)w(P).$$

By Lemma 4.4 and Corollary 4.2,

$$C(T) \leq 2n \sum_{v \in V} d_G(v, R) + (n^2/2)w(R)$$
$$\leq 2n \sum_{v \in V} d_{\widehat{T}}(v, P) + (2n^2/3)w(P)$$
$$\leq (3/2)C(\widehat{T}).$$

☐

THEOREM 4.3
There is a 3/2-approximation algorithm with time complexity $O(n^4)$ for the MRCT problem.

PROOF First, the all-pairs shortest paths can be found in $O(n^2 \log n + mn)$. For every triple (r_1, r_0, r_2) of vertices, we construct $R = SP_G(r_1, r_0) \cup SP_G(r_0, r_2)$ and $T \in \text{star}(R)$ including the degenerated cases $r_i = r_j$. By Lemma 4.9, at least one of these stars is a 3/2-approximation solution of the MRCT problem, and we can choose the one with the minimum routing cost. For the time complexity, we show that each star can be constructed in $O(n)$ time. By Lemma 4.3, a $T \in \text{star}(R)$ can be constructed in $O(n)$ time if for every $v \in V$, a shortest path from v to R is given. Define $A(i, j, k) = d_G(i, SP_G(j, k))$ and $B(i, j, k)$ to be the vertex in $SP_G(j, k)$ which is closest to i. It is easy to see that $A(i, j, k)$ and $B(i, j, k)$ can be computed in $O(n^4)$ time.[1] For any $R = SP_G(r_1, r_0) \cup SP_G(r_0, r_2)$, since

$$d_G(v, R) = \min\{A(v, r_1, r_0), A(v, r_0, r_2)\},$$

$d_G(v, R)$ as well as the vertex in R closest to v can be computed in total $O(n^4)$ time for all $v \in V$ and for all such R at a preprocessing step. Finally, for any spanning tree T, we can compute $C(T)$ in $O(n)$ time. So the total time complexity is $O(n^4)$. ☐

---

[1] Remark: It can be computed in $O(n^3)$ time by dynamic programming. However the total time complexity is still $O(n^4)$.

4.3.4 Further improvement

Let S be a minimal δ-separator of $\widehat{T}$. The strategy of algorithms shown in this section is to "guess" the structure of S and to construct a general star with the guessed structure as the core. If $T \in \text{star}(S)$, by Lemmas 4.4 and 4.5,

$$C(T) \leq 2n \sum_{v \in V(G)} d_G(v, S) + (n^2/2)w(S),$$

and

$$C(\widehat{T}) \geq 2(1-\delta)n \sum_{v \in V} d_{\widehat{T}}(v, S) + 2\delta(1-\delta)n^2 w(S).$$

The approximation ratio, by comparing the two inequalities, is

$$\max\{\frac{1}{1-\delta}, \frac{1}{4\delta(1-\delta)}\}.$$

The ratio achieves its minimum when the two terms coincide, i.e., $\delta = 1/4$, and the minimum ratio is $4/3$. In fact, by using a general star and a $(1/4)$-separator, it is possible to approximate an MRCT with ratio $(4/3) + \varepsilon$ for any constant $\varepsilon > 0$ in polynomial time. The additional error ε is due to the difference between the guessed and the true separators.

By this strategy, the approximation ratio is limited even if S was known exactly. The limit of the approximation ratio may be mostly due to that we consider only general stars. In a general star, the vertices are always connected to their closest vertices of the core. In extreme cases, roughly half of the vertices connected are to both sides of a costly edge. This results in the cost $(n^2/2)w(S)$ in the upper bound of a general star. To make a breakthrough, the restriction that each vertex is connected to the closest vertex of the core needs to be relaxed.

A metric graph is a complete graph with triangle inequality, i.e., each edge is a shortest path of its two endpoints. Recall the proof of Lemma 4.9. If the input graph is a metric graph, the core will be a two-edge path, and each vertex is adjacent to one of the "critical vertices"— a centroid and the two endpoints of a path separator. Define k-stars to be the trees with at most k internal vertices. The constructed approximation solution is a 3-star. More importantly, k-stars have no such restriction like general stars and can be used to approximate an MRCT more precisely. Later in this chapter we shall see how it works.

However, k-stars work only for metric graphs. The class of metric graphs is an important subclass of graphs. Solving the MRCT problem on metric graphs is itself meaningful. Before considering the approximation problem on metric graphs, two questions come to our minds:

- What is the computational complexity of the MRCT problem on metric graphs, NP-hard or polynomial-time solvable?

- If it is NP-hard, does its approximability differ from that of the general problem?

We shall answer the questions in the next section.

4.4 A Reduction to the Metric Case

In this section, we shall show that the MRCT problem on general inputs can be reduced to the same problem with metric inputs. The reduction is done by a transformation algorithm.

DEFINITION 4.6   *The metric closure of a graph $G = (V, E, w)$ is the complete graph $\bar{G} = (V, V \times V, \bar{w})$ in which $\bar{w}(u,v) = d_G(u,v)$ for all $u, v \in V$.*

Let $G = (V, E, w)$ and $\bar{G} = (V, V \times V, \bar{w})$ be its metric closure. Any edge (a,b) in $\bar{G}$ is called a *bad edge* if $(a,b) \notin E$ or $w(a,b) > \bar{w}(a,b)$. For any bad edge $e = (a,b)$, there must exist a path $P = SP_G(a,b) \neq e$ such that $w(P) = \bar{w}(a,b)$. Given any spanning tree T of $\bar{G}$, the algorithm can construct another spanning tree Y without any bad edge such that $C(Y) \leq C(T)$. Since Y has no bad edge, $\bar{w}(e) = w(e)$ for all $e \in E(Y)$, and Y can be thought of as a spanning tree of G with the same routing cost. The algorithm is listed in the following.

Algorithm: REMOVE_BAD
Input: A spanning tree T of $\bar{G}$.
Output: A spanning tree Y of G such that $C(Y) \leq C(T)$.
        Compute all-pairs shortest paths of G.
(I)   **while** there exists a bad edge in T
        Pick a bad edge (a,b). Root T at a.
        /* assume $SP_G(a,b) = (a, x, ..., b)$ and y is the parent of x */
        if b is not an ancestor of x **then**
            $Y^* \leftarrow T \cup (x,b) - (a,b); Y^{**} \leftarrow Y^* \cup \{(a,x)\} - \{(x,y)\}$;
        else
            $Y^* \leftarrow T \cup (a,x) - (a,b); Y^{**} \leftarrow Y^* \cup \{(b,x)\} - \{(x,y)\}$;
        if $C(Y^*) < C(Y^{**})$ **then** $Y \leftarrow Y^*$ **else** $Y \leftarrow Y^{**}$
(II) $T \leftarrow Y$

The algorithm computes Y by iteratively replacing the bad edges until there is no bad edge. It will be shown that the cost is never increased at each iteration and it takes no more than $O(n^2)$ iterations. We assume that

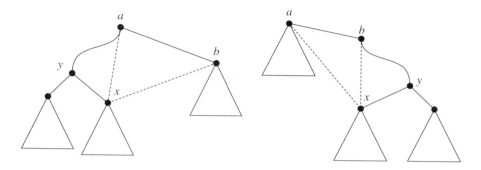

FIGURE 4.6: Remove bad edge (a, b). Case 1 (left) and Case 2 (right).

the shortest paths obtained in the first step have the following property: If $SP_G(a,b) = (a, x, \ldots, b)$, then $SP_G(a,b) = (a, x) \cup SP_G(x,b)$. This assumption is not strong since almost all popular algorithms for all-pairs shortest paths output such a solution.

PROPOSITION 4.1
The while loop in Algorithm REMOVE_BAD *is executed at most $O(n^2)$ times.*

PROOF For each bad edge $e = (a, b)$, let $h(e)$ be the number of edges in $SP_G(a, b)$ and $f(T) = \sum_{\text{bad } e} h(e)$. Since $h(e) \leq n - 1$, $f(T) < n^2$ initially. Since (a, x) is not a bad edge, it is easy to check that $f(T)$ decreases by at least 1 at each iteration. □

PROPOSITION 4.2
Before instruction **(II)** *is executed, $C(Y) \leq C(T)$.*

PROOF For any node v, let $S_v = V(T_v)$. As shown in Figure 4.6, there are two cases. Case 2 is identical to Case 1 if the tree is re-rooted at b and the roles of a and b are exchanged. Therefore, only the inequality for Case 1 needs to be proved, i.e., $x \in S_a - S_b$.

If $C(Y^*) \leq C(T)$, the result follows. Otherwise, let $U_1 = S_a - S_b$ and $U_2 = S_a - S_b - S_x$. Since the distance does not change for any two vertices both in U_1 (or both in S_b), we have

$$C(T) < C(Y^*)$$
$$\Rightarrow \sum_{u \in U_1} \sum_{v \in S_b} d_T(u, v) < \sum_{u \in U_1} \sum_{v \in S_b} d_{Y^*}(u, v).$$

Since for all $u \in U_1$ and $v \in S_b$, $d_T(u,v) = d_T(u,a) + \bar{w}(a,b) + d_T(b,v)$ and $d_{Y^*}(u,v) = d_T(u,x) + \bar{w}(x,b) + d_T(b,v)$,

$$\sum_{u \in U_1} \sum_{v \in S_b} (d_T(u,a) + \bar{w}(a,b) + d_T(b,v))$$
$$< \sum_{u \in U_1} \sum_{v \in S_b} (d_T(u,x) + \bar{w}(x,b) + d_T(b,v))$$
$$\Rightarrow |S_b| \sum_{u \in U_1} d_T(u,a) + |U_1||S_b|\bar{w}(a,b)$$
$$< |S_b| \sum_{u \in U_1} d_T(u,x) + |U_1||S_b|\bar{w}(x,b)$$
$$\Rightarrow \sum_{u \in U_1} d_T(u,a) + |U_1|\bar{w}(a,b) < \sum_{u \in U_1} d_T(u,x) + |U_1|\bar{w}(x,b).$$

Note that $S_b \neq \emptyset$ since the inequality holds. By the definition of the metric closure, $\bar{w}(a,b) = \bar{w}(a,x) + \bar{w}(x,b)$, and then

$$\sum_{u \in U_1} (d_T(u,a) - d_T(u,x)) < -|U_1|\bar{w}(a,x). \tag{4.13}$$

Now consider the cost of Y^{**}.

$$(C(Y^{**}) - C(T))/2 = \sum_{u \in U_2} \sum_{v \in S_x} (d_{Y^{**}}(u,v) - d_T(u,v))$$
$$+ \sum_{u \in U_1} \sum_{v \in S_b} (d_{Y^{**}}(u,v) - d_T(u,v)).$$

Since $d_{Y^{**}}(u,v) \leq d_T(u,v)$ for $u \in U_1$ and $v \in S_b$, the second term is not positive. By observing that $d_T(u,v) = d_T(u,x) + d_T(x,v)$ and $d_{Y^{**}}(u,v) = d_T(u,a) + \bar{w}(a,x) + d_T(x,v)$ for any $u \in U_2$ and $v \in S_x$, we obtain

$$(C(Y^{**}) - C(T))/2$$
$$\leq \sum_{u \in U_2} \sum_{v \in S_x} (d_T(u,a) + \bar{w}(a,x) - d_T(u,x))$$
$$= |S_x| \sum_{u \in U_2} (d_T(u,a) + \bar{w}(a,x) - d_T(u,x))$$
$$= |S_x| \sum_{u \in U_2} (d_T(u,a) - d_T(u,x)) + |U_2||S_x|\bar{w}(a,x)$$
$$\leq |S_x| \sum_{u \in U_1} (d_T(u,a) - d_T(u,x)) + |U_2||S_x|\bar{w}(a,x) \tag{4.14}$$
$$< -|U_1||S_x|\bar{w}(a,x) + |U_2||S_x|\bar{w}(a,x) \tag{4.15}$$
$$\leq 0.$$

(4.14) is obtained by observing that $U_1 - U_2 = S_x$ and $d_T(u,a) > d_T(u,x)$ for any $u \in S_x$. (4.15) is derived by applying (4.13). Therefore, $C(Y^{**}) < C(T)$ and the result follows. □

The next lemma follows Propositions 4.1 and 4.2, and that each iteration can be done in $O(n)$ time.

LEMMA 4.10

For any spanning tree $\bar{T}$ of $\bar{G}$, it can be transformed into a spanning tree T of G in $O(n^3)$ time and $C(T) \leq C(\bar{T})$.

The above lemma implies that $C(\mathrm{mrct}(G)) \leq C(\mathrm{mrct}(\bar{G}))$. It is easy to see that $C(\mathrm{mrct}(G)) \geq C(\mathrm{mrct}(\bar{G}))$. Therefore, we have the following corollary.

COROLLARY 4.3

$C(\mathrm{mrct}(G)) = C(\mathrm{mrct}(\bar{G}))$.

Let ΔMRCT denote the MRCT problem with metric inputs. We have the next theorem.

THEOREM 4.4

If there is a $(1+\varepsilon)$-approximation algorithm for $\Delta MRCT$ with time complexity $O(f(n))$, then there is a $(1+\varepsilon)$-approximation algorithm for MRCT with time complexity $O(f(n) + n^3)$.

PROOF Let G be the input graph for the MRCT problem. The metric closure $\bar{G}$ can be constructed in time $O(n^2 \log n + mn)$. If there is a $(1+\varepsilon)$-approximation algorithm for the ΔMRCT problem, a spanning tree T of $\bar{G}$ can be computed in time $O(f(n))$ such that $C(T) \leq (1+\varepsilon)C(\mathrm{mrct}(\bar{G}))$. Using Algorithm REMOVE_BAD, a spanning tree Y of G can be constructed such that

$$C(Y) \leq C(T) \leq (1+\varepsilon)C(\mathrm{mrct}(\bar{G})) = (1+\varepsilon)C(\mathrm{mrct}(G)).$$

The overall time complexity is then $O(f(n) + n^3)$. □

COROLLARY 4.4

The $\Delta MRCT$ problem is NP-hard.

4.5 A Polynomial Time Approximation Scheme

4.5.1 Overview

We present a *Polynomial Time Approximation Scheme* (PTAS) for the MRCT problem in this section. The main result is the following theorem.

THEOREM 4.5
There is a PTAS for finding a minimum routing cost tree of a weighted undirected graph. Specifically, we can find a $(1+\varepsilon)$-approximation solution in time $O(n^{2\lceil \frac{2}{\varepsilon} \rceil - 2})$.

As described previously, the fact that the costs w may not obey the triangle inequality is irrelevant, since we can simply replace these costs by their metric closure. Therefore, in this section we may assume that $G = (V, E, w)$ is a metric graph.

We use k-stars, i.e., trees with no more than k internal nodes, as a basis of our approximation scheme. In Section 4.5.5 we show that for any constant k, a minimum routing cost k-star can be determined in polynomial time. In order to show that a k-star achieves a $(1 + \varepsilon)$ approximation, we show that, for any tree T and constant $\delta \leq 1/2$:

1. It is possible to determine a δ-separator, and the separator can be cut into several δ-paths such that the total number of cut nodes and leaves of the separator is at most $\lceil \frac{2}{\delta} \rceil - 3$. (Lemma 4.11)

2. Using the separator, T can be converted into a $(\lceil \frac{2}{\delta} \rceil - 3)$-star X, whose internal nodes are just those cut nodes and leaves. The routing cost of X satisfies $C(X) \leq (1 + \frac{\delta}{1-\delta})C(T)$. (Lemma 4.13)

By using $T = \widehat{T} = \text{mrct}(G)$, $\delta = \frac{\varepsilon}{1+\varepsilon}$ and finding the best $(\lceil \frac{2}{\delta} \rceil - 3)$-star K, we obtain the desired approximation

$$C(K) \leq (1 + \frac{\delta}{1-\delta})C(\widehat{T}) = (1+\varepsilon)C(\widehat{T}).$$

Before going into the details of the general case, take a look at how to find a $3/2$-approximation of an MRCT and its performance analysis.

Recall that a centroid of a tree is the vertex whose removal cuts the tree into components of no more than $n/2$ vertices. Let $\widehat{T}$ be an MRCT of a metric graph G. Root $\widehat{T}$ at its centroid r. For an edge (u,v) with parent v, the routing load of the edge is $2x(n-x)$, in which x is the number of descendants of u. (For convenience, we assume that a vertex is also a descendant of itself.) For a desired positive $\delta \leq 1/2$, removing all vertices with number of descendants no more than δn, we may obtain a connected subgraph S of T. The subgraph S is a minimal δ-separator. In the case that $\delta = 0.5$, S contains only the centroid. If the δ-separator S of the MRCT is given and, for each vertex not in S, its lowest ancestor in S is also known, we may easily construct a $1/(1-\delta)$-approximation Y of the MRCT as follows:

- $S \subset Y$.

- For each vertex u not in S, connect it to its lowest ancestor v in S by adding edge (u,v).

The approximation ratio of Y can be shown by the following two observations.

- For each edge in S, the routing load in Y is the same as that in $\widehat{T}$.

- For each edge (u,v) not in S, the routing load in Y is $2(n-1)$. However, there is a path from u to v in $\widehat{T}$ of which each edge has routing load no less than $2(1-\delta)n$, and the length of the path is at least the same as the edge (u,v).

For example, consider the simplest case that $\delta = 1/2$. A $1/2$-separator contains only one vertex. Assume that r be such a vertex on an MRCT $\widehat{T}$. For every other vertex, the ancestor in the separator is r. Connecting all other nodes to r, we obtain a star Y. The routing cost of Y is the sum of all edges (v,r) multiplied by its routing load $2(n-1)$. For each vertex v, there is a path from v to r in $\widehat{T}$. Since r is a $1/2$-separator of $\widehat{T}$, the routing load of the path is at least n and the path is no shorter than the edge (v,r) by the triangle inequality. Consequently Y is a 2-approximation of the MRCT $\widehat{T}$. Since the separator contains only one vertex, we may try all possible vertices and it leads to an $O(n^2)$ time 2-approximation of the MRCT problem on metric graphs.

But when $\delta < 1/2$, the separator may contain as many as $\Omega(n)$ vertices, and it is too costly to enumerate all possible separators. However, instead of the entire separator, we may achieve the same ratio by knowing only some *critical* vertices of the separator. In the following, we shall take $\delta = 1/3$ as an example to show that we only need to know three critical vertices of the separator to construct a $3/2$-approximation.

Root the MRCT $\widehat{T}$ at its centroid r. There are at most two subtrees which contain more than $n/3$ nodes. Let a and b be the lowest vertices with at least $n/3$ descendants. We ignore the cases where $r = a$ or $r = b$. For such cases, the ratio can be shown similarly. Let P be the path from a to b. Obviously the path contains r and is a minimal $1/3$-separator. We shall transform T into a 3-star with internal nodes a, b, and r such that its routing cost is no more than $3/2$ of that of $\widehat{T}$. As shown in Figure 4.7, let us partition all the vertices into V_a, V_b, V_r, V_{ar}, and V_{br}. The sets V_a, V_b, and V_r contain the

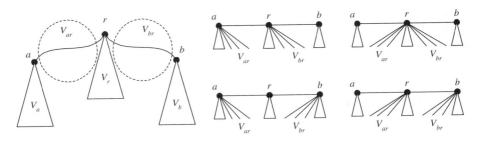

FIGURE 4.7: An MRCT and the four 3-stars.

vertices whose lowest ancestor on P is a, b, and r, respectively. The set V_{ar} (and V_{br}) consists of the vertices whose lowest ancestor on P is between a and r (and between b and r, respectively).

First replace P with edges (a,r) and (b,r). For each vertex v in V_a (or V_b, V_r), edge (v,a) (or (v,b), (v,r) respectively) is added. For the vertices in V_{ar}, we consider two cases. Either all of them are connected to a or all of them are connected to r. The vertices in V_{br} are connected similarly. The four possible 3-stars are illustrated in Figure 4.7.

Now consider the routing cost of $\widehat{T}$. For each vertex v, the routing load of the path from v to P is no less than $4n/3$ since P is a 1/3-separator. For each edge e of P, since there are at least $n/3$ nodes on either side of it, the routing load is no less than $2(n/3)(2n/3) = 4n^2/9$. Therefore we have the following lower bound of the routing cost of the MRCT:

$$C(\widehat{T}) \geq (4n/3) \sum_{v} d_{\widehat{T}}(v, P) + (4/9)n^2 w(P).$$

In either of the 3-stars constructed above, for each vertex v in $V_a \cup V_r \cup V_b$, the routing load of the edge incident to v is $2(n-1)$, and the edge length is at most the same as the path from v to P on T. For each node v in V_{ar}, by the triangle inequality, we have

$$(w(v,a) + w(v,r))/2 \leq d_{\widehat{T}}(v, P) + d_{\widehat{T}}(a,r)/2.$$

Note that there are no more than $n/6$ nodes in V_{ar}. For edge (a,r), the routing load is no more than $2(n/2)(n/2) = n^2/2$. (Note: For this simple case that $\delta = 1/3$, this bound is enough. However, a more precise analysis of the incremental routing load is required for smaller δ.)

For the nodes in V_{br}, we may obtain a similar result. In summary, the minimum routing cost of the constructed 3-stars is no more than

$$2(n-1)\sum_v d_{\widehat{T}}(v,P) + (n^2/6)(d_{\widehat{T}}(a,r) + d_{\widehat{T}}(b,r)) + (1/2)n^2 w(P)$$

$$\leq 2n \sum_v d_{\widehat{T}}(v,P) + (2/3)n^2 w(P).$$

The approximation ratio, by comparing with the lower bound of the optimal, is 3/2.

The method can be extended to any $\delta \leq 0.5$. Let S be a δ-separator of T. The critical vertex set, defined as the *cut and leaf set* in the next section, to construct a $1/(1-\delta)$-approximation k-star consists of the following vertices.

- The leaves of S as a and b in the above example.

- The vertices with more than two neighbors on S.

- Some additional vertices such that all the critical vertices cut the separator into edge-disjoint paths and the number of vertices whose lowest ancestors on S belong to the same path is no more than $\delta n/2$. Such a path is defined as a δ-path in the next section. In the above example, r is such a vertex. The vertices a, b and r cut the separator into two paths, and the number of vertices in either V_{ar} or V_{br} is no more than $n/6$.

We shall show that the number of the necessary critical vertices is at most $2/\delta - 3$ for any $\delta \leq 0.5$. Consequently there exists a $(2/\delta - 3)$-star which is an approximation of an MRCT with ratio $1/(1-\delta)$. The PTAS is to construct the $(2/\delta - 3)$-star of minimum routing cost.

The core of a tree is the subgraph obtained by removing all its leaves. The core of a k-star contains no more than k vertices and therefore the number of all possible cores is polynomial to k. For each possible core, the algorithm finds the best way to connect the leaves to one of the vertices of the core. A k-component integer vector $(n_1, n_2, \ldots, n_k)$ is used to indicate how many leaves will be connected to each of the k vertices of the core, in which $\sum_i n_i = n - k$. There are $O(n^{k-1})$ such vectors. For each core and each vector, the routing load on each core edge is fixed since the number of vertices on both sides of the edge are specified by the core and the vector. Therefore the best leaf connection is determined by the leaf edges subject to the numbers of leaves to be connected to the vertices of the core. Such a problem can be solved by solving an assignment problem in $O(n^3)$ time. Consequently the minimum routing cost k-star can be constructed in time polynomial to k and n.

Consider an example for $k = 3$. For a 3-star, the core is a 3-vertex path. There are $O(n^3)$ possible cores. For each possible core (a, b, c), use a three-component integer vector (x, y, z) to indicate how many leaves will be connected to a, b, and c, in which $x + y + z = n - 3$ and x, y, z are nonnegative integers. There are $O(n^2)$ such vectors. For a specified core and a specified

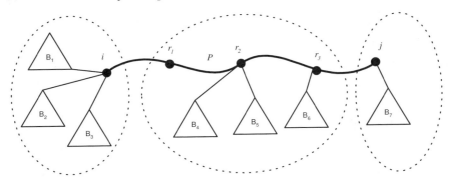

FIGURE 4.8: $B_1, \ldots, B_7$ are branches of P. $VB(T, P, i) = \{i\} \cup V(B_1) \cup V(B_2) \cup V(B_3)$. P^c is the number of vertices in $\{r_1, r_2, r_3\} \cup V(B_4) \cup V(B_5) \cup V(B_6)$.

vector, the routing load on each core edge is also fixed. That is, the routing load on (a, b) is $2(x+1)(n-x-1)$ and on (b, c) is $2(z+1)(n-z-1)$. Therefore the best leaf connection is determined by the leaf edges subject to the numbers of leaves to be connected to a, b, c. Such a problem can be solved by solving an assignment problem in $O(n^3)$ time. The total time complexity is $O(n^{2k+2})$, which is polynomial for constant k. In fact we need not solve the assignment problem for the best leaf connection of each vector. The best leaf connection of one vector can be found from that of another vector by solving a shortest path problem if the two vectors are adjacent. Two vectors are adjacent if one can be obtained from the other by increasing a component by one and decreasing a component by one, e.g., $(5, 4, 2)$ and $(5, 3, 3)$. With this result, the time complexity is reduced to $O(n^{2k})$.

4.5.2 The δ-spine of a tree

Let $P = SP_T(i, j)$ in which $|VB(T, P, i)| \geq |VB(T, P, j)|$. We shall use the following notations to simplify the expressions.

- $P^a = |VB(T, P, i)|$.
- $P^b = |VB(T, P, j)|$.
- $P^c = n - |VB(T, P, i)| - |VB(T, P, j)|$.
- $Q(P) = \sum_{1 \leq x \leq h} |VB(T, P, r_x)| \times d_T(r_x, i)$, where $P = (i, r_1, r_2, \ldots, r_h, j)$.

P^a and P^b are the numbers of vertices that are hanging at the two end points of the path. Note that we always assume $P^a \geq P^b$. In the case that P contains only one edge, $P^c = 0$. The notations are illustrated in Figure 4.8.

DEFINITION 4.7 Let $1 \leq k \leq n$. A k-star is a spanning tree of G which has no more than k internal nodes. An optimal k-star is the k-star with the minimum routing cost.

We now turn to the notions of δ-paths and δ-spines. Informally, a δ-path is a path such that not too many nodes (at most $\delta n/2$) are hanging at its internal nodes. A δ-spine is a set of edge-disjoint δ-paths, whose union is a minimal δ-separator. That is, a δ-spine is obtained by cutting a minimal δ-separator into δ-paths. In the case that the minimal δ-separator contains just one node, the only δ-spine is the empty set.

DEFINITION 4.8 Given a spanning tree T of G, and $0 < \delta \leq 0.5$, a δ-path of T is a path P such that $P^c \leq \delta n/2$.

DEFINITION 4.9 Let $0 < \delta \leq 0.5$. A δ-spine $Y = \{P_1, P_2, ..., P_h\}$ of T is a set of pairwise edge-disjoint δ-paths in T such that $S = \bigcup_{1 \leq i \leq h} P_i$ is a minimal δ-separator of T. Furthermore, for any pair of distinct paths P_i and P_j in the spine, we require that either they do not intersect or, if they do, the intersection point is an endpoint of both paths.

DEFINITION 4.10 Let Y be a δ-spine of a tree T. $CAL(Y)$ (which stands for the cut and leaf set of Y) is the set of the endpoints of the paths in Y. In the case that Y is empty, the cut and leaf set contains only one node which is a δ-separator of T. Formally $CAL(Y) = \{u, v | \exists P = SP_T(u, v) \in Y\}$ if Y is not empty, and otherwise $CAL(Y) = \{u | u \text{ is a minimal } \delta\text{-separator }\}$.

Example 4.4

In Figure 4.9(a), S (bold lines) is a minimal 1/4 separator of the tree. Vertex v_1 is a centroid, and vertices v_2, v_3, and v_4 are leaves in S. In (b), the separator is cut into a 1/4-spine of the tree. The CAL of the spine is $\{v_1, v_2, v_3, v_4, v_5\}$. The path between v_1 and v_4 is cut at vertex v_5 to ensure that the number of vertices hung at each path is no more than $n/8$. □

LEMMA 4.11

For any constant $0 < \delta \leq 0.5$, and a spanning tree T of G, there exists a δ-spine Y of T such that $|CAL(Y)| \leq \lceil 2/\delta \rceil - 3$.

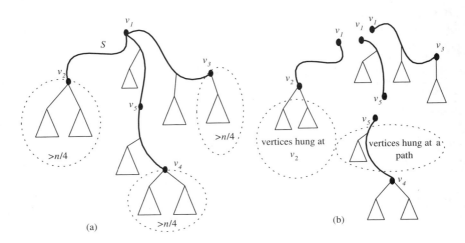

FIGURE 4.9: Separator, spine and CAL. Each triangle represents a subtree with no more than $n/4$ vertices.

PROOF Let S be a minimal δ-separator of T. S is a tree. Let U_1 be the set of leaves in S, U_2 be the set of vertices which have more than two neighbors in S, and $U = U_1 \cup U_2$. Let $h = |U_1|$. Clearly, $|U| \leq 2h - 2$. Let Y_1 be the set of paths obtained by cutting S at all the vertices in U_2. For example, for the tree on the right side of Figure 4.9, $U_1 = \{v_2, v_3, v_4\}$; $U_2 = \{v_1\}$; Y_1 contains $SP_T(v_1, v_2)$, $SP_T(v_1, v_3)$ and $SP_T(v_1, v_4)$. For any $P \in Y_1$, if $P^c > \delta n/2$ then P is called a *heavy* path. It is easy to check that Y_1 satisfies the requirements of a δ-spine except that there may exist some heavy paths. Suppose P is not a δ-path. We can break it up into δ-paths by the following process. First find the longest prefix of P starting at one of its endpoints and ending at some internal vertex, say i, in the path, that determines a δ-path. Now we break P at vertex i. Then we repeat the breaking process on the remaining suffix of P starting at i, stripping off the next δ-path and so on. In this way P can be cut into δ-paths by breaking it up at no more than $\lceil 2P^c/(\delta n) \rceil - 1$ vertices. Since there are at least δn nodes hung at each leaf,

$$\sum_{P \in Y_1} P^c < n - h\delta n.$$

Let U_3 be the minimal vertex set for cutting the heavy paths to result in a δ-spine Y of T. We have

$$|U_3| \leq \lceil 2(n - h\delta n)/(\delta n) \rceil - 1 = \lceil 2/\delta \rceil - 2h - 1.$$

So, $|CAL(Y)| = |U| + |U_3| \leq \lceil 2/\delta \rceil - 3$. □

Example 4.5
For any tree, there always exist a (1/3)-spine and a (1/4)-spine whose cut and leaf set contains no more than 3 and 5 vertices, respectively, as illustrated in Figure 4.9. Taking $\delta = 1/5$, it follows that there exists a (1/5)-spine whose cut and leaf set has no more than 7 vertices. □

4.5.3 Lower bound

We are going to establish a more precise lower bound than Lemma 4.5 (Section 4.3). While proving Lemma 4.5, we substituted $l(T,e)$ by $2\delta(1-\delta)n^2$ in (4.7),

$$C(T) \geq 2(1-\delta)n \sum_{v \in V} d_T(v,S) + \sum_{e \in E(S)} l(T,e)w(e).$$

Now we give a more careful analysis of the last term. Let Y be a δ-spine of a spanning tree T of G and $S = \bigcup_{P \in Y} P$ be a minimal δ-separator of T. Rewriting in terms of δ-spine and recalling $l(T,e) = 2e^a e^b$, we have

$$\sum_{e \in E(S)} l(T,e)w(e) = 2 \sum_{P \in Y} \sum_{e \in P} e^a e^b w(e).$$

Assume $P = (r_0, r_1, r_2, \ldots, r_h)$ in which $|VB(T,P,r_0)| \geq |VB(T,P,r_h)|$. Let $|VB(T,P,r_i)| = n_i$ for $1 \leq i \leq h-1$ and $e_i = (r_{i-1}, r_i)$ for $1 \leq i \leq h$.

$$\sum_{e \in P} e^a e^b w(e)$$

$$= \sum_{i=1}^{h} \left(P^a + P^c - \sum_{j=i}^{h-1} n_j \right) \left(P^b + \sum_{j=i}^{h-1} n_j \right) w(e_i)$$

$$\geq \sum_{i=1}^{h} P^b (P^a + P^c) w(e_i) + (P^a - P^b) \sum_{i=1}^{h} \sum_{j=i}^{h-1} n_j w(e_i)$$

$$+ \sum_{i=1}^{h} \left(\sum_{j=i}^{h-1} n_j \right) \left(P^c - \sum_{j=i}^{h-1} n_j \right) w(e_i)$$

$$\geq P^b(P^a + P^c)w(P) + (P^a - P^b) \sum_{j=1}^{h-1} n_j \left(\sum_{i=1}^{j} w(e_i) \right)$$

$$= P^b (P^a + P^c) w(P) + (P^a - P^b) Q(P).$$

Note that $Q(P)$ is defined at the first paragraph of Section 4.5.2. From this and (4.7), we obtain the next lemma.

LEMMA 4.12
Let Y be a δ-spine of a spanning tree T of G and $S = \bigcup_{P \in Y} P$ be a minimal

δ-separator of T. Then

$$C(T) \geq 2(1-\delta)n \sum_{v \in V} d_T(v, S) + 2 \sum_{P \in Y} \left(P^b(P^a + P^c)w(P) + (P^a - P^b)Q(P) \right).$$

4.5.4 From trees to stars

Let $\delta \leq 1/2$ and $k = \lceil 2/\delta \rceil - 3$. We now decompose an optimal solution $\widehat{T}$ and construct a k-star whose routing cost is upper bounded by $\frac{k+3}{k+1}C(\widehat{T})$.

Let $Y = \{P_i | 1 \leq i \leq h\}$ be a δ-spine of $\widehat{T}$ in which $|CAL(Y)| \leq \lceil 2/\delta \rceil - 3$. Note that the set of all the edges in Y forms a δ-separator S. Assume $P_i = SP_{\widehat{T}}(u_i, v_i)$ and $|VB(\widehat{T}, P_i, u_i)| \geq |VB(\widehat{T}, P_i, v_i)|$. By the following steps, we construct a k-star whose internal nodes are exactly the cut and leaf set of the δ-spine we just identified.

1. We connect these nodes by short-cutting edges along the spine to construct a tree spanning these nodes with the same skeletal structure as the spine.

2. All vertices in subtrees hanging at the cut and leaf nodes of the spine are connected directly to their closest node in the spine.

3. Along a δ-path in the spine, all the internal nodes and nodes in subtrees hanging at internal nodes are connected to one of the two endpoints of this path (note that both are in the cut and leaf set of the spine) in such a way as to minimize the resulting routing cost.

This is the k-star used to argue the upper bound on the routing cost.

Example 4.6
In Figure 4.10, we illustrate how to construct the desired k-star from an optimal tree. Frame (a) is an optimal tree in which the separator is shown and the cut and leaf set is $\{A, B, C, D, E\}$. Frame (b) is the tree spanning the cut and leaf nodes, which has the same skeletal structure as the spine. Frames (c), (d) and (e) illustrate how to connect other nodes to the cut and leaf nodes. Frame (c) exhibits the nodes hanging at a δ-path. These nodes will be connected as in either Frame (d) or (e). The nodes hanging at the endpoints of the path will be connected to the endpoints in either case. All the internal nodes of the path and nodes hanging at the internal nodes will be connected to one of the two endpoints. Note that they are connected to the same endpoint either as Frame (d) or Frame (e), but not connected to the two endpoints partially. □

More formally, construct a subgraph $R \subset G$ with vertex set $CAL(Y)$ and edge set $E_r = \{(u_i, v_i) | 1 \leq i \leq h\}$. Trivially, R is a tree. Let $f(i)$ be an

Minimum Routing Cost Spanning Trees

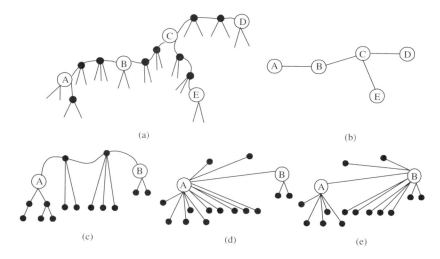

FIGURE 4.10: Constructing the k-star from an optimal tree.

indicator variable such that

$$f(i) = \begin{cases} 1 & \text{if } (P_i^a - P_i^b)\, P_i^c w(P_i) \geq n\,(2Q(P_i) - P_i^c w(P_i)) \\ 0 & \text{otherwise} \end{cases}$$

The indicator variable $f(i)$ determines the endpoint of P_i to which all the internal nodes and nodes hanging at such internal nodes will be directly connected. We construct a spanning tree X of G where the edge set E_x is determined by the following rules:

1. $R \subset X$.

2. If $q \in VB(\widehat{T}, S, r)$ then $(q, r) \in E_x$, for any $r \in \{u_i, v_i | 1 \leq i \leq h\}$.

3. For the vertex set $V_i = V - VB(\widehat{T}, P_i, u_i) - VB(\widehat{T}, P_i, v_i)$, if $f(i) = 1$ then $\{(q, u_i) | q \in V_i\} \subset E_x$, else $\{(q, v_i) | q \in V_i\} \subset E_x$. That is, the vertices in V_i are either all connected to u_i or all connected to v_i.

PROPOSITION 4.3
X is a k-star.

The above proposition is trivial; consider the cost of X.

PROPOSITION 4.4
$C(X) \leq \frac{1}{1-\delta} C(\widehat{T})$.

PROOF Since any edge in $E_x - E_r$ is incident with a leaf,

$$\frac{C(X)}{2} = \sum_{e \in E_r} e^a e^b w(e) + (n-1) \sum_{e \in E_x - E_r} w(e).$$

First, for any $e = (u_i, v_i) \in E_r$,

$$e^a e^b w(e) \leq \left(P_i^a + f(i) P_i^c\right) \left(P_i^b + (1 - f(i)) P_i^c\right) w(P_i)$$
$$= P_i^a P_i^b w(P_i) + \left(f(i) P_i^b + (1 - f(i)) P_i^a\right) P_i^c w(P_i).$$

Second, by the triangle inequality and recalling that for subgraph $S \subset \widehat{T}$, $d_{\widehat{T}}^S(i,j)$ stands for $w(SP_{\widehat{T}}(i,j) \cap S)$, we have

$$\sum_{e \in E_x - E_r} w(e) \leq \sum_{v \in V} d_{\widehat{T}}(v, S) + \sum_{i=1}^h \sum_{v \in V_i} \left(f(i) d_{\widehat{T}}^S(v, u_i) + (1 - f(i)) d_{\widehat{T}}^S(v, v_i)\right)$$
$$= \sum_{v \in V} d_{\widehat{T}}(v, S) + \sum_{i=1}^h \left(f(i) Q(P_i) + (1 - f(i)) \left(P_i^c w(P_i) - Q(P_i)\right)\right).$$

Thus,

$$\frac{C(X)}{2} \leq \sum_{i=1}^h P_i^a P_i^b w(P_i) + n \sum_{v \in V} d_{\widehat{T}}(v, S)$$
$$+ \sum_{i=1}^h \min\{P_i^b P_i^c w(P_i) + n Q(P_i), P_i^a P_i^c w(P_i) + n (P_i^c w(P_i) - Q(P_i))\}.$$

Since the minimum of two numbers is not larger than their weighted mean, we have

$$\min\{P_i^b P_i^c w(P_i) + n Q(P_i), P_i^a P_i^c w(P_i) + n \left(P_i^c w(P_i) - Q(P_i)\right)\}$$
$$\leq \left(P_i^b P_i^c w(P_i) + n Q(P_i)\right) \frac{P_i^a}{P_i^a + P_i^b}$$
$$+ \left(P_i^a P_i^c w(P_i) + n \left(P_i^c w(P_i) - Q(P_i)\right)\right) \frac{P_i^b}{P_i^a + P_i^b}.$$

Then,

$$\frac{C(X)}{2} \leq \sum_{i=1}^{h} P_i^a P_i^b w(P_i) + n \sum_{v \in V} d_{\widehat{T}}(v, S) + \sum_{i=1}^{h} \frac{(2P_i^a P_i^b P_i^c + n P_i^b P_i^c) w(P_i)}{P_i^a + P_i^b}$$

$$+ \sum_{i=1}^{h} \frac{(P_i^a - P_i^b) n Q(P_i)}{P_i^a + P_i^b}$$

$$= n \sum_{v \in V} d_{\widehat{T}}(v, S) + \sum_{i=1}^{h} \frac{w(P_i)}{P_i^a + P_i^b} \left((P_i^a P_i^b + P_i^b P_i^c) n + P_i^a P_i^b P_i^c \right)$$

$$+ \sum_{i=1}^{h} \frac{(P_i^a - P_i^b) n Q(P_i)}{P_i^a + P_i^b}.$$

The simplification in the last inequality uses the observation that for any i, we have $P_i^a + P_i^b + P_i^c = n$. By Lemma 4.12,

$$C(X) \leq C(\widehat{T}) \max_{1 \leq i \leq h} \left\{ \frac{1}{1-\delta}, \frac{n}{P_i^a + P_i^b} + \frac{P_i^a P_i^c}{(P_i^a + P_i^b)(P_i^a + P_i^c)} \right\}.$$

Since $P_i^c \leq \delta n/2$,

$$\frac{n}{P_i^a + P_i^b} + \frac{P_i^a P_i^c}{(P_i^a + P_i^b)(P_i^a + P_i^c)}$$

$$\leq \frac{n}{P_i^a + P_i^b} + \frac{P_i^c}{P_i^a + P_i^b}$$

$$= \frac{n + P_i^c}{n - P_i^c} \leq \frac{2+\delta}{2-\delta} \leq \frac{1}{1-\delta}.$$

This completes the proof. □

For any integer $k \geq 1$, we take $\delta = \frac{2}{k+3}$. By the above two propositions, we have the next lemma.

LEMMA 4.13
An optimal k-star of a metric graph is a $(k+3)/(k+1)$ approximation of an MRCT.

Example 4.7
Taking $\delta = 1/3$ in Lemma 4.13, it follows that there exists a 3-star which is a 1.5-approximation of an MRCT, which coincides with the result stated in Section 4.3.4. Taking $\delta = 0.2$, it follows that there exists a 7-star which is a 1.25-approximation. □

In the following section we will show that it is possible to determine an optimal k-star of a graph in polynomial time. In fact, we have the following lemma.

LEMMA 4.14
An optimal k-star of a graph G can be constructed in $O(n^{2k})$ time.

The proof is delayed to the next section. The following theorem establishes the time-complexity of our PTAS.

THEOREM 4.6
There exists a PTAS for the ΔMRCT problem, which can find a $(1+\varepsilon)$-approximation solution in $O(n^\rho)$ time complexity where $\rho = 2\lceil 2/\varepsilon \rceil - 2$.

PROOF By Lemma 4.13, there exists a k-star which is a $(k+3)/(k+1)$ approximation of an MRCT. For finding a $(1+\varepsilon)$-approximation solution, we take $k = \lceil 2/\varepsilon \rceil - 1$ and find an optimal k-star. The time complexity is $O(n^\rho)$ where $\rho = 2\lceil 2/\varepsilon \rceil - 2$ from Lemma 4.14. □

The result in Theorem 4.5 is immediately derived from Theorems 4.6 and 4.4.

4.5.5 Finding an optimal k-star

In this section we describe an algorithm for finding an optimal k-star in G for a given value of k. As mentioned before, given an accuracy parameter $\varepsilon > 0$, we apply this algorithm for $k = \lceil \frac{2}{\varepsilon} - 1 \rceil$, and return an optimal k-star as a $(1+\varepsilon)$-approximate solution.

For a given k, to find an optimal k-star, we consider all possible subsets S of vertices of size k, and for each such choice, find an optimal k-star where the remaining vertices have degree one.

4.5.5.1 A polynomial-time method

First, we verify that the overall complexity of this step is polynomially bounded for any fixed k. Any k-star can be described by a triple $(S, \tau, \mathcal{L})$, where $S = \{v_1, \ldots, v_k\} \subseteq V$ is the set of k distinguished vertices which may have degree more than one, τ is a spanning tree topology on S, and $\mathcal{L} = (L_1, \ldots, L_k)$, where $L_i \subseteq V - S$ is the set of vertices connected to vertex $v_i \in S$. For any $r \in Z^+$, an r-vector is an integer vector with r components. Let $l = (l_1, \ldots, l_k)$ be a nonnegative k-vector such that $\sum_{i=1}^k l_i = n - k$. We say that a k-star $(S, \tau, \mathcal{L})$ has the configuration (S, τ, l) if $l_i = |L_i|$ for all $1 \le i \le k$.

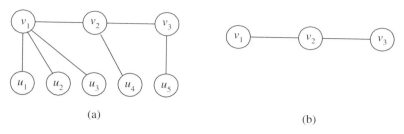

FIGURE 4.11: The configuration of a 3-star.

Example 4.8
For the 3-star shown in Figure 4.11(a), $S = \{v_1, v_2, v_3\}$, τ is shown in (b), $L_1 = \{u_1, u_2, u_3\}$, $L_2 = \{u_4\}$, $L_3 = \{u_5\}$, and $l = (3, 1, 1)$. □

For a fixed k, the total number of configurations is $O(n^{2k-1})$ since there are $\binom{n}{k}$ choices for S, k^{k-2} possible tree topologies on k vertices, and $\binom{n-1}{k-1}$ possible such k-vectors. (To see this, observe that every such vector can be put in correspondence with picking $k-1$ among $n-1$ linearly ordered elements, and using the cardinalities of the segments between consecutively picked segments as the components of the vector.) Note that any two k-stars with the same configuration have the same routing load on their corresponding edges. We define $\alpha(S, \tau, l)$ to be an optimal k-star with configuration (S, τ, l).

Note that any vertex v in $V - S$ that is connected to a node $s \in S$ contributes a term of $w(v, s)$ multiplied by its routing load of $2(n-1)$. Since all these routing loads are the same, the best way of connecting the vertices in $V - S$ to nodes in S is obtained by finding a minimum-cost way of matching up the nodes of $V - S$ to those in S which obeys the degree constraints on the nodes of S imposed by the configuration, and the costs are the distances w. This problem can be solved in polynomial time for a given configuration by a straightforward reduction to an instance of minimum-cost perfect matching. The above minimum-cost perfect matching problem, also called the *assignment* problem, has been well studied and can be solved in $O(n^3)$ time (cf. [1]). Therefore, the overall complexity is $O(n^{2k+2})$ for finding an optimal k-star.

4.5.5.2 A faster method

In the PTAS, we need to solve many matching problems, each for one configuration. It takes polynomial time to solve these problems individually and this result is sufficient for showing the existence of the PTAS. Although there is no obvious way to reduce the time complexity for one matching problem, the total time complexity can be significantly reduced when considering all these matching problems together. The key point is that two optimal k-stars of similar configurations share a large common portion in their structures. By carefully ordering the matching problems for the configurations and exploiting the common structure of two consecutive problems, we can obtain an optimal solution of any configuration in this order by performing a single augmentation on the optimal solution of the previous configuration. Thus, we show (Lemma 4.15) how to compute $\alpha(S, \tau, l)$ for a given configuration in $O(nk)$ time.

Let W_{ab} be the set of all nonnegative a-vectors whose entries add up to a constant b. In $W_{ab} \times W_{ab}$, we introduce the relation $\sim$ as $l \sim l'$ if there exist $1 \leq s, t \leq a$ such that

$$l'_i = \begin{cases} l_i - 1 & \text{if } i = s \\ l_i + 1 & \text{if } i = t \\ l_i & \text{otherwise} \end{cases}$$

For a pair l and l' such as the above, we say that l' *is obtained from l by s and t.*

Let $r = |W_{ab}| = \binom{a+b-1}{a-1}$. The following proposition shows that the elements of W_{ab} can be linearly ordered as $l^1, \ldots, l^r$ so that $l^{i+1} \sim l^i$ for all $1 \leq i \leq r-1$.

PROPOSITION 4.5
For all positive integers a, b, there exists a permutation $\pi^{a,b}$ of W_{ab} such that $\pi_1^{a,b}$ is the lexicographic minimum, $\pi_r^{a,b}$ is the lexicographic maximum, and $\pi_{i+1}^{a,b} \sim \pi_i^{a,b}$ for all $i = 1, \ldots, r-1$.

PROOF By induction. The claim is clearly true when $a = 1$ for any b. Assume the claim is true for all b when $a = m - 1$. For $a = m$ construct the ordering as follows. First the elements for which $l_1 = 0$ ordered applying $\pi^{a-1,b}$ to $(l_2, \ldots, l_a)$. Then the elements for which $l_1 = 1$, ordered according to *decreasing* $\pi^{a-1,b-1}$. In general each block for which $l_1 = h$ is ordered by applying $\pi^{a-1,b-h}$ to $(l_2, \ldots, l_a)$, forward or backwards according to the parity of h. Note that $\pi_{i+1}^{a,b} \sim \pi_i^{a,b}$ within one block. Furthermore, at block boundaries the part $(l_2, \ldots, l_a)$ is either a lexicographic minimum or maximum so that it is feasible to increase by one l_1. Finally, it is obvious that the first and the last of the constructed ordering are the lexicographic minimum and maximum respectively. □

Example 4.9
The ordering of $W_{3,4}$ is as follows:

$$(0,0,4)\ (0,1,3)\ (0,2,2)\ (0,3,1)\ (0,4,0)$$
$$(1,3,0)\ (1,2,1)\ (1,1,2)\ (1,0,3)\ (2,0,2)$$
$$(2,1,1)\ (2,2,0)\ (3,1,0)\ (3,0,1)\ (4,0,0)$$

According to Proposition 4.5 we can order the elements of $W_{k,(n-k)}$ as $l^1, \ldots, l^r$, where $r = \binom{n-1}{k-1}$. Note that $l^1 = (0, \ldots, 0, n-k)$ and $l^r = (n-k, 0, \ldots, 0)$. We shall show how to obtain $\alpha(S, \tau, l^{i+1})$ from $\alpha(S, \tau, l^i)$.

LEMMA 4.15
$\alpha(S, \tau, l^{i+1})$ can be computed from $\alpha(S, \tau, l^i)$ in $O(nk)$ time.

PROOF We shall show that $\alpha(S, \tau, l^{i+1})$ can be found from $\alpha(S, \tau, l^i)$ by means of a shortest path computation. A similar argument is used for solving a minimum cost flow problem given the solution of another minimum cost flow problem which differs by only one unit capacity arc (see Exercise 10.20 in [1]).

For convenience, let us rename the vertices so that $S = \{1, \ldots, k\}$. Let $l^i = (|L_1|, \ldots, |L_k|)$ and $(S, \tau, \mathcal{L}) = \alpha(S, \tau, l^i)$. Let us define an auxiliary weighted digraph $D(\mathcal{L}) = (V, A, \delta)$ in which the arc set is

$$A = \{(u,v) | u \in V - S, v \in S\} \cup \{(u,v) | u \in S, v \in L_u\},$$

and $\delta(u,v) = w(u,v)$ if $u \notin S$, and $\delta(u,v) = -w(u,v)$ if $u \in S$. For a node in S, the weight on an outgoing arc reflects the cost reduction for removing a leaf from its neighbors, and the weight on an incoming arc reflects the increase in cost for connecting a leaf to the node.

Any cycle (not necessarily simple) in the graph describes a way of changing $(S, \tau, \mathcal{L})$ into another k-star with the same configuration, and the difference in cost between the new and the old k-stars is given by $2(n-1)$ times the length of the cycle. Since $(S, \tau, \mathcal{L})$ is optimal for its configuration, we conclude that there is no negative length cycle in $D(\mathcal{L})$. Consequently any shortest path between two vertices must be a simple path.

Similarly, if l^{i+1} is obtained from l^i by s and t, then any path from s to t in the auxiliary graph changes $(S, \tau, \mathcal{L})$ into a k-star with configuration (S, τ, l^{i+1}). Conversely, any k-star with configuration (S, τ, l^{i+1}) can be obtained by a path (not necessarily simple) from s to t. Since the contributed cost of a path is proportional to its length, it is clear that there is a shortest path from s to t, changing $\alpha(S, \tau, l^i)$ into $\alpha(S, \tau, l^{i+1})$. We now show how such a shortest path can be computed in $O(kn)$ time.

Consider any shortest path $(u_1, v_1, u_2, \ldots, v_{h-1}, u_h)$ between two nodes $u_i \in S$ and $v_i \in V - S$ in $D(\mathcal{L})$. Take two consecutive edges (u_i, v_i) and (v_i, u_{i+1}) in the path. Since the path is shortest, v_i must be such as to minimize the sum of the two edge lengths. Recall that $\delta(u_i, v_i) = -w(u_i, v_i)$ and $\delta(v_i, u_{i+1}) = w(v_i, u_{i+1})$. Then, we have that the sum of the two edge lengths is $\min_{v_i \in L_i} \{w(v_i, u_{i+1}) - w(u_i, v_i)\}$. To find the shortest path from s to t on $D(\mathcal{L})$, we construct a complete digraph $D'(\mathcal{L})$ with vertex set S and lengths

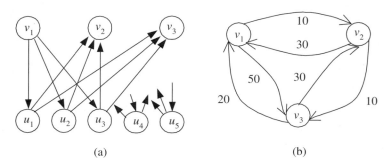

FIGURE 4.12: The auxiliary graphs for the faster method.

δ', in which
$$\delta'(i,j) = \min_{v \in L_i}\{w(v,j) - w(i,v)\}.$$

It is easy to see that the length of the shortest path from s to t on $D'(\mathcal{L})$ is the same as the one on $D(\mathcal{L})$.

Given the graph $D'(\mathcal{L})$, a shortest s-t path (and also the corresponding path on $D(\mathcal{L})$) can be found in $O(k^2)$ time. Finally, to construct $D'(\mathcal{L})$, for each vertex $i \in S$, we have to find $k-1$ minima (one for each other $j \in S$), each over a set of l_i elements. Adding up, the total time complexity is

$$(k-1)\sum_{i=1}^{k} l_i = (k-1)(n-k) = O(nk).$$

☐

Example 4.10

Suppose that the edge lengths $w(v_i, u_j)$ in Figure 4.11 are given in the following matrix.

$$\begin{array}{c} \\ v_1 \\ v_2 \\ v_3 \end{array} \begin{array}{c} u_1 \ u_2 \ u_3 \ u_4 \ u_5 \\ \begin{bmatrix} 10 & 10 & 30 & 50 & 50 \\ 50 & 60 & 40 & 20 & 60 \\ 80 & 70 & 80 & 30 & 30 \end{bmatrix} \end{array}$$

The 3-star in Figure 4.11(a) is optimal for $l = (3, 1, 1)$ and we want to compute an optimal for vector $(2, 1, 2)$ which is obtained from l by v_1 and v_3. The auxiliary digraph $D(\mathcal{L})$ is shown in Figure 4.12(a) and $D'(\mathcal{L})$ is shown in (b). Since $L_1 = \{u_1, u_2, u_3\}$, the length $\delta'(v_1, v_2)$ is obtained by

$$\min \left\{ \begin{array}{c} w(v_2, u_1) - w(v_1, u_1) \\ w(v_2, u_2) - w(v_1, u_2) \\ w(v_2, u_3) - w(v_1, u_3) \end{array} \right\} = w(v_2, u_3) - w(v_1, u_3) = 10,$$

which corresponds to the cost of moving u_3 from L_1 to L_2. The shortest path from v_1 to v_3 in $D'(\mathcal{L})$ is (v_1, v_2, v_3) and has length 20. Thus the best leaf connection of configuration $(S, \tau, (2, 1, 2))$ is $L_1 = \{u_1, u_2\}$, $L_2 = \{u_3\}$ and $L_3 = \{u_4, u_5\}$. □

We are now able to prove Lemma 4.14 which states that an optimal k-star can be found in $O(n^{2k})$ time.

PROOF When S and τ are fixed, to find an optimal k-star we begin by $\alpha(S, \tau, l^1)$, which is readily obtained by setting $L_k = V - S$. Then, using Lemma 4.15, we compute optimal k-stars for configurations $l^2, \ldots, l^r$, and report the best overall.

Since we have $\binom{n}{k}$ choices for S, k^{k-2} possible tree topologies, and for each fixed S and τ, we have $\binom{n-1}{k-1}$ configurations, where an optimal for each configuration can be computed in $O(nk)$ time, therefore, in total we can find an optimal k-star in $O(n^k) \times O(n^{k-1}) \times O(n) = O(n^{2k})$ time. □

4.6 Applications

4.6.1 Network design

An obvious application of an MRCT is in network design. The length of an edge reflects the cost of routing along the edge. The input graph is the underlying graph in which the edges represent all possible links between nodes. One may want to construct a fixed network such that message or cargo may be transported on the network from sources to destinations.

Trees are an important network structure because of the following two features:

1. Trees are connected subgraphs with minimum number of edges.

2. The routing algorithm on a tree is very simple.

An MRCT is a spanning tree minimizing the total routing cost when the communication requirements for all pairs of vertices are equal. Or in the stochastic model, an MRCT is a spanning tree having the minimum expected routing cost if the requirement between any pair of vertices has equal probability.

4.6.2 Computational biology

Besides the obvious connection to network design, trees with small routing cost also find application in the construction of good multiple sequence alignments in computational biology. To explicate the application, we shall first briefly introduce the concept of multiple sequence alignments. More details can be found in textbooks for computational biology such as [49].

4.6.2.1 Multiple sequence alignments

Multiple sequence alignments are important tools for finding patterns common to a set of genetic sequences in computational biology. A multiple alignment of a set of n strings involves inserting gaps (blanks) in the strings and arranging their characters into columns with n rows, one from each string. The order of characters along a row corresponding to string s_i is the same as that in s_i, with possibly some blanks inserted. The following is an example of an alignment of three strings TCCGATG, CCGGACG and TCGACG.

T	C	C	-	G	A	T	-	G
-	C	C	G	G	A	-	C	G
T	C	-	-	G	A	-	C	G

The intent of identifying common patterns is represented by attempting as much as possible to place the same character in every column.

The multiple sequence alignment problem has typically been formalized as an optimization problem in which some explicit objective function is minimized or maximized. One of the most popular objective functions for multiple alignment generalizes ideas from alignment of two sequences. The pairwise-alignment problem [91] can be phrased as that of finding a minimum mutation path between two sequences. Formally, given costs for inserting or deleting a character and for substituting one character of the alphabet for another, the problem is to find a minimum-cost mutation path from one sequence to the other. The cost of this path is known as the *edit distance* in computer science. An optimal alignment of two sequences of length l can be computed effectively by dynamic programming [91] in $O(l^2)$ time. The generalization to multiple sequences leads to the sum-of-pairs objective.

The *sum-of-pairs* or SP objective for multiple alignment is to minimize the sum, over all pairs of sequences, of the pairwise distance between them *in the alignment* (where the distance of two sequences in an alignment with l columns is obtained by adding up the costs of the pairs of characters appearing at positions $1, \ldots, l$).

Pioneering work of Sankoff and co-authors [83] led to an exponential-time dynamic programming solution to the SP-alignment problem. A straightforward implementation requires time proportional to $2^n l^n$, for a problem with n sequences each of length at most l. Considering that in typical real-life instances l can be a few hundred, the basic dynamic programming approach turns out to be infeasible for all but very small problems.

4.6.2.2 Approximation algorithms via routing cost trees

The first approximation algorithm for the SP-alignment problem was due to Gusfield [48] with a performance ratio of $2 - \frac{2}{n}$ where n is the number of sequences aligned. This was slightly improved to $2 - \frac{3}{n}$ by Pevzner [77]. The best known approximation algorithm for this problem is due to Bafna, Lawler and Pevzner [6] which achieves a ratio of $2 - \frac{r}{n}$ for any fixed value of r. The running time is exponential in r. Notice that this is not a PTAS for the problem, and no polynomial time approximation scheme is known yet for the SP-alignment problem.

Gusfield's approximation algorithm for the SP-alignment problem is based on the 2-approximation for minimum routing cost trees due to Wong [92]. The algorithm uses a folklore approach to multiple alignment guided by a tree, due to Feng and Doolittle [34]: Given a spanning tree on the complete graph on the sequences to be aligned, the multiple alignment guided by the tree is built recursively as follows. First, remove a leaf sequence l in the tree attached to sequence v by a tree edge (l, v), and align the remaining sequences recursively. Then, insert back the leaf sequence in the alignment guided by an optimal pairwise alignment between the pair l and v. If this optimal pairwise alignment introduces a gap in v, insert the same gap in the recursively computed alignment for the tree without the leaf. Since the cost of aligning a blank to a blank is assumed zero, the resulting alignment has the property that for every pair related by a tree edge, the cost of the induced pairwise alignment equals to their edit distance. By the triangle inequality on edit-distances, the SP-cost of the alignment derived from this spanning tree is upper-bounded by the routing cost of the tree.

Wong's 2-approximation algorithm is the one introduced in Section 4.2. For graphs with metric distances obeying the triangle inequality, every shortest-paths tree is isomorphic to a star. Furthermore, in this case, Wong's analysis shows that the best star has routing cost at most twice the total cost of the graph itself. The cost of the graph in this case is the sum of pairwise edit distances between sequences, which is a lower bound on the SP-cost. Thus, Gusfield observed that a multiple alignment derived from the best center-star gives a 2-approximation for the SP-alignment problem.

4.6.2.3 Tree-driven SP-alignment

Despite the popularity of the SP-objective, most of the currently available methods for finding alignments use a *progressive* approach of incrementally building the alignment adding sequences one at a time with no performance guarantee on the SP-cost. The Feng-Doolittle procedure can be viewed as one such procedure. The advantages of such approaches is their low running time, but the shortcoming is that the order in which the sequences are merged into the alignment determines its cost.

In trying to define a middle ground between the SP-objective and the more practical progressive methods, the tree-driven SP-alignment method was proposed: apply the Feng-Doolittle procedure to the *best possible* spanning tree in the complete graph on the sequences. By the reasoning above, the tree that gives the best upper bound on the SP-cost of the alignment is the one with the minimum routing cost. Thus, the PTAS for routing cost trees may be useful in finding good trees for applying any progressive alignment method such as the Feng-Doolittle procedure.

4.7 Summary

In this chapter, we investigate how to approximate a minimum routing cost spanning tree of a graph. The first algorithm approximates an MRCT by constructing a shortest-paths tree rooted at some vertex. By comparing with a trivial lower bound, the cost is shown within two times the optimal. By the technique of solution decomposition, we can have another proof of the approximation ratio. More importantly the technique can be extended to designing better approximation algorithms. It is exhibited by approximation algorithms with ratio 15/8 and 3/2, which use general stars instead of shortest-paths trees to approximate an MRCT.

However, a difficulty is encountered while attempting better approximation ratios for general graphs. A transformation algorithm is introduced to resolve the difficulty. The algorithm REMOVE_BAD can construct a spanning tree of a graph from a spanning tree of its metric closure without increasing the routing cost. The transformation not only shows the NP-hardness of the MRCT problem on metric graphs but also reduces the approximation problem on general graphs to the metric case.

By decomposing an optimal solution, it is shown that there exists a k-star which is a good approximation of an MRCT of a metric graph. The approximation ratio approaches to 1 as k is increased. A PTAS is then presented by showing how to construct an optimal k-star in polynomial time for fixed k.

Finally the applications of an MRCT in network design and in computational biology are discussed.

Bibliographic Notes and Further Reading

Te Chiang Hu [55] formulated a general version of the routing cost spanning tree problem that he called optimum communication spanning trees (OCT), cf. [ND7] in [43]. In this problem, in addition to the costs on edges, a requirement value λ_{ij} is specified for every pair of vertices i, j. The communication cost between a pair in a spanning tree is the cost of the path between them in the tree multiplied by their requirement λ_{ij}. Thus the routing cost is a special case of the communication cost when all the requirement values are one. He used the term "optimum distance spanning trees" to denote trees with a minimum routing cost and derived a weak condition under which the optimum routing cost tree is a star.

David S. Johnson et al. [60] showed that the MRCT problem on a general graph is NP-hard in the strong sense. Robert Dionne et al. [29] studied the exact and the heuristic algorithms. The MRCT problem is also known by the name *shortest total path length spanning tree problem*, cf. [ND3] in [43]. R. Wong [92] showed that there exists a shortest-paths tree which is a 2-approximation of an MRCT and gave a worst-case analysis of the approximation algorithm. Bang Ye Wu, Kun-Mao Chao, and Chuan Yi Tang [97] made a breakthrough on the approximation ratio by using the general stars and giving a lower bound with the solution decomposition technique. They showed that it is possible to approximate an MRCT with ratio $(4/3) + \varepsilon$ for any positive constant ε in polynomial time. The idea was soon generalized to the PTAS by Bang Ye Wu et al. [100]. The term "MRCT" first appeared in the paper. Matteo Fischetti et al. [35] studied the techniques for finding the exact solution while avoiding an exhaustive search.

Several extensions of the MRCT problem will be discussed in the next chapter, including the OCT problem, the sum-requirement OCT problem and the product-requirement OCT problem. Some variants where not all vertices are sources are also included.

More details of the multiple sequence alignment as well as other problems in computational biology can be found in [49].

Exercises

4-1. What is an MRCT of a complete graph with unit length on each edge? Prove your answer.

4-2. What is an MRCT of a complete bipartite graph $K_{m,n}$ with unit length on each edge? Prove your answer.

4-3. What is the routing cost of an n-vertex path with unit cost on each edge?

4-4. Show that the problem of finding a minimum routing cost path visiting each vertex exactly once is NP-hard. (Hint: Consider the Hamiltonian path problem.)

4-5. Design an algorithm for finding a centroid of a tree. What is the time complexity of your algorithm?

4-6. Give a tree with two centroids.

4-7. Let $r: V \to Z^+$ be a vertex weight and define the r-centroid of a tree T to be the vertex c such that if we remove c from T, the total vertex weight of each component is no more than half of the total vertex weight. Generalize the algorithm in Exercise 4-5 to find a r-centroid of a tree.

4-8. Show that for a tree with positive edge lengths, the median coincides with the centroid.

4-9. Design an algorithm to find a δ-separator of a tree. What is the time complexity?

4-10. Design an algorithm to find a δ-spine of a tree. What is the time complexity?

4-11. Let Y be the output of the algorithm REMOVE_BAD with input T. Show that $w(Y) \leq w(T)$.

4-12. In the definition of the MRCT problem we assume the cost of an edge is symmetric, i.e., $w(u,v) = w(v,u)$. Make a discussion for the MRCT problem with asymmetric edge cost.

4-13. Let $S = \{1, 2, 3, 4\}$ be a vertex set. How many tree topologies are there on S? Enumerate all possible tree topologies.

4-14. Write down the ordering of W_{ab} in Proposition 4.5 for (a) $a = 2$ and $b = 5$; (b) $a = 4$ and $b = 3$.

4-15. Design an algorithm for generating the ordering of W_{ab} in Proposition 4.5.

4-16. What is the sum-of-pair (SP) score of the alignment shown in Section 4.6.2.1? Assume that the scores of a match and a mismatch are 1 and -1, respectively.

4-17. A subsequence is obtained by removing some characters from a sequence. A sequence is a *longest common subsequence* (LCS) of two sequences if it is a subsequence of both sequences and its length (number of characters) is maximum. Find a dynamic programming algorithm for finding an LCS of two sequences in the literature. (You can find it in most of the textbooks about algorithms.)

4-18. Suppose that each insertion and deletion takes unit cost. The edit distance between two sequences is the minimum number of operations (insertion or deletion) to transform a sequence into the other. What is the relation between the edit distance and the length of an LCS?

Chapter 5

Optimal Communication Spanning Trees

5.1 Introduction

The *optimal communication spanning tree* (OCT) problem is defined as follows. Let $G = (V, E, w)$ be an undirected graph with nonnegative edge length function w. We are also given the requirements $\lambda(u,v)$ for each pair of vertices. For any spanning tree T of G, the communication cost between two vertices is defined to be the requirement multiplied by the path length of the two vertices on T, and the communication cost of T is the total communication cost summed over all pairs of vertices. Our goal is to construct a spanning tree with minimum communication cost. That is, we want to find a spanning tree T such that $\sum_{u,v \in V} \lambda(u,v) d_T(u,v)$ is minimized.

The requirements in the OCT problem are arbitrary nonnegative values. By restricting the requirements, several special cases of the problem were defined in the literature. We list the problems in the following, in which $r: V \to Z_0^+$ is a given vertex weight function and $S \subset V$ is a set of k vertices given as sources.

- $\lambda(u,v) = 1$ for each $u, v \in V$: This version is the MINIMUM ROUTING COST SPANNING TREE (MRCT) problem discussed in the previous chapter.

- $\lambda(u,v) = r(u)r(v)$ for each $u, v \in V$: This version is called the OPTIMAL PRODUCT-REQUIREMENT COMMUNICATION SPANNING TREE (PROCT) problem.

- $\lambda(u,v) = r(u) + r(v)$ for each $u, v \in V$: This version is called the OPTIMAL SUM-REQUIREMENT COMMUNICATION SPANNING TREE (SROCT) problem.

- $\lambda(u,v) = 0$ if $u \notin S$: This version is called the p-SOURCE OCT (p-OCT) problem. In other words, the goal is to find a spanning tree minimizing $\sum_{u \in S} \sum_{v \in V} \lambda(u,v) d_T(u,v)$.

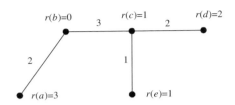

FIGURE 5.1: The product-requirement communication cost between vertices a and d is $3 \times 2 \times (2+3+2) = 42$, and the sum-requirement communication cost between vertices a and e is $(3+1) \times (2+3+1) = 24$.

- $\lambda(u,v) = 1$ if $u \in S$, and $\lambda(u,v) = 0$ otherwise: This version is called the p-Source MRCT (p-MRCT) problem. In other words, the goal is to find a spanning tree minimizing $\sum_{u \in S} \sum_{v \in V} d_T(u,v)$.

We define two communication costs and notations for the PROCT and the SROCT problems.

DEFINITION 5.1 *The* product-requirement communication *(or p.r.c. in abbreviation) cost of a tree T is defined by $C_p(T) = \sum_{u,v} r(u)r(v)d_T(u,v)$.*

When there are more than one vertex weight functions, we shall use $C_p(T, r)$ to indicate that the cost is with respect to weight r.

DEFINITION 5.2 *The* sum-requirement communication *(or s.r.c. in abbreviation) cost of a tree T is defined by $C_s(T) = \sum_{u,v}(r(u)+r(v))d_T(u,v)$.*

Given a graph G, the PROCT (or SROCT) problem asks for a spanning tree T of G such that $C_p(T)$ (or $C_s(T)$ respectively) is minimum among all possible spanning trees.

Example 5.1
The p.r.c. cost and s.r.c. cost between a pair of vertices are illustrated in Figure 5.1. The cost of the tree is the sum of the cost for all pairs of vertices.
☐

The relationship of the different versions of the OCT problems is illustrated in Figure 5.2. Note that there are variants for the multi-source problems. By "arbitrary p," we mean there is no restriction on the number of sources in the input data, while by "fixed p," the number of sources is always equal to the constant p.

In this chapter, we shall discuss the computational complexities and algorithms for these problems.

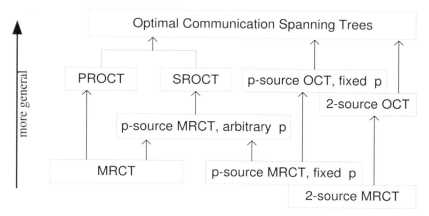

FIGURE 5.2: The relationship of the OCT problems.

5.2 Product-Requirement

A formal definition of the problem is in the following.

PROBLEM: Optimal Product-Requirement Communication Spanning Trees (PROCT).
INSTANCE: A graph $G = (V, E, w)$ with vertex weight $r : V \to Z_0^+$.
GOAL: Find a spanning tree T of minimum p.r.c. cost.

The PROCT problem is reduced to the MRCT problem if all vertices have the same weight. Thus PROCT problem is NP-hard. In this section, we shall focus on how to approximate a PROCT. For a vertex set U, we use $r(U)$ to denote $\sum_{u \in U} r(u)$, and $r(H) = r(V(H))$ for a graph H. Let $R = r(G)$ denote the total vertex weight of the input graph.

5.2.1 Overview

Recall that the PTAS for the MRCT problem in Chapter 4 is obtained by showing the following properties:

1. The MRCT problem on general graphs is equivalent to the problem on metric graphs.

2. A k-star is a spanning tree with at most k internal nodes. The minimum routing cost k-star is a $((k+3)/(k+1))$-approximation solution for the metric MRCT problem.

3. For a fixed k, a minimum routing cost k-star on a metric can be found in polynomial time.

The PROCT problem is a weighted counterpart of the MRCT problem. A vertex with weight $r(v)$ can be regarded as a super node consisting of $r(v)$ nodes of unit weight and connected by edges of zero length. In fact, the first and the second properties remain true for the PROCT problem. They can be obtained by straightforward generalizations of the previous results. Consequently, a polynomial time algorithm for a minimum p.r.c. cost k-star is a PTAS for the PROCT problem. However, there is no obvious way to generalize the algorithm for the minimum routing cost k-star to that for the minimum p.r.c. cost k-star. A straightforward generalization conducts to a pseudo-polynomial time algorithm whose time complexity depends on the total weight of all vertices.

In this section, we first check some properties derived from their unweighted counterpart in Chapter 4. Then two approximation algorithms are presented. The first one approximates a PROCT by finding the minimum p.r.c. cost 2-star. By the previous result, such a 2-star is a 5/3-approximation of a PROCT. But with a more precise analysis, it can be shown that it is in fact a 1.577-approximation solution. The second algorithm in this section is a PTAS, which employs a PTAS for a minimum p.r.c. cost k-star.

5.2.2 Preliminaries

5.2.2.1 Centroid and p.r.c. routing loads

Similar to a centroid of an unweighted graph, we define the r-centroid of a tree with vertex weight function r.

DEFINITION 5.3 *Let T be a tree with vertex weight function r. The r-centroid of a tree T is a vertex $m \in V(T)$ such that if we remove m, then $r(H) \leq r(T)/2$ for any branch H.*

Example 5.2
Both the centroid and the r-centroid of the tree in Figure 5.1 are vertex c. If $r(b) = 2$ instead of zero, the r-centroid will be vertex b. □

The definitions of the p.r.c. routing load and the p.r.c. routing cost on an edge are also similar to their unweighted counterparts.

DEFINITION 5.4 *Let T be any spanning tree of a graph G, and r a vertex weight function. For any edge $e = (u, v) \in E(T)$, we define the p.r.c. routing load on the edge e to be $l_p(T, r, e) = 2r(T_u)r(T_v)$, where T_u and T_v are the two subgraphs obtained by removing e from T. The p.r.c. routing cost on the edge e is defined to be $l_p(T, r, e)w(e)$.*

Similar to its unweighted counterpart, the p.r.c. routing cost can also be computed by summing up the edge lengths multiplied by their p.r.c. routing load. We state it as the next lemma, and the proof is left as an exercise.

LEMMA 5.1
Let T be any spanning tree of a graph $G = (V, E, w)$ and r be a vertex weight function. $C_p(T, r) = \sum_{e \in E(T)} l_p(T, r, e) w(e)$.

Example 5.3
Let T be the tree in Figure 5.1. The p.r.c. load of edge (b, c) is $2(3 + 0)(1 + 2 + 1) = 24$ and the p.r.c. cost of T can be computed as follows:

$$C_p(T, r) = 2 \times 3 \times 4 \times w(a, b) + 2 \times 3 \times 4 \times w(b, c) + 2 \times 5 \times 2 \times w(c, d)$$
$$+ 2 \times 6 \times 1 \times w(c, e)$$
$$= 172.$$

□

5.2.2.2 Reduction to metric graphs

In Chapter 4, it is shown that the MRCT problem on a general graph can be reduced to the problem on its metric closure. The reduction is by Algorithm REMOVE_BAD which constructs a spanning tree of G from a spanning tree of its metric closure without increasing the routing cost. By a straightforward generalization, it can be shown that the algorithm also works for the p.r.c. routing cost. We state the results in the following but omit the proofs.

LEMMA 5.2
Given a spanning tree T of $\bar{G}$, there is an algorithm which can construct a spanning tree Y of G with $C_p(Y) \leq C_p(T)$ in $O(n^3)$ time.

Let proct(G) denote an optimal solution of the PROCT problem with input graph G. The above lemma implies that

$$C_p(\text{proct}(G)) \leq C_p(\text{proct}(\bar{G})).$$

It is easy to see that

$$C_p(\text{proct}(G)) \geq C_p(\text{proct}(\bar{G})).$$

Therefore, we have the following corollary.

COROLLARY 5.1
$C_p(\text{proct}(G)) = C_p(\text{proct}(\bar{G}))$.

COROLLARY 5.2
If there is a $(1+\varepsilon)$-approximation algorithm for $\Delta PROCT$ with time complexity $O(f(n))$, then there is a $(1+\varepsilon)$-approximation algorithm for $PROCT$ with time complexity $O(f(n)+n^3)$.

5.2.2.3 Balanced k-stars

In Chapter 4, it is shown that there exists a k-star which is a good approximation of the MRCT. Similarly we can show the property for the PROCT problem. However, for the convenience of showing the PTAS in Section 5.2.4, we define a *balanced k-star*. The additional restriction does not affect the property.

DEFINITION 5.5 *Let k be a positive integer. A balanced k-star is a spanning tree with at most k internal nodes and its core is a minimal $(2/(k+3))$-separator of the spanning tree.*

The core of a k-star is the subgraph obtained by removing all its leaves. Similar to Lemma 4.13, we have the next result.

LEMMA 5.3
There exists a balanced k-star of $\bar{G}$, which is a $(k+3)/(k+1)$-approximation of a PROCT of $\bar{G}$.

Essentially, Lemma 5.3 is a refinement of Lemma 4.13, in which it is shown that there is a k-star X which is a $(k+3)/(k+1)$-approximation solution for the MRCT problem. The k-star X is actually a balanced k-star. The adjustment in the definition is for the convenience of showing the approximation ratio. The lemma can be proved by just replacing the vertex cardinalities with the total weight of the corresponding vertex sets in all relevant definitions and lemmas.

In the remaining paragraphs of this section, an optimal balanced k-star is a balanced k-star of minimum p.r.c. cost.

5.2.2.4 Approximating a PROCT

By Corollary 5.2 and Lemma 5.3, we can only focus on the problem of finding an optimal balanced k-star on a metric graph.

> PROBLEM: Optimal Balanced k-Stars
> INSTANCE: metric graph $G = (V, E, w)$ with vertex weight $r : V \to Z^+$ and an integer k.
> GOAL: Find an optimal balanced k-star.

When k is a constant, similar to the algorithm of finding a minimum routing cost k-star, we may try all possible k-vertex trees as the core since there are at most $\binom{n}{k}k^{k-2}$ possible cores. For each core, we need to solve the following subproblem.

> PROBLEM: Optimal Balanced k-Stars with a Given Core
> INSTANCE: A metric graph $G = (V, E, w)$, a tree A in G with $|V(A)| = k$, and a vertex weight $r : V \to Z^+$.
> GOAL: Find an optimal balanced k-star with core A if it exists.

As a result, we can reduce the problem of approximating a PROCT to the problem of finding (or approximating) an optimal balanced k-star with a given core.

LEMMA 5.4
For any fixed integer k, if there exists an algorithm for finding an optimal balanced k-star with a given core in $O(f(n))$ time, there exists an approximation algorithm for a PROCT with ratio $(k+3)/(k+1)$ and time complexity $O(n^k f(n) + n^3)$.

COROLLARY 5.3
For any fixed integer k, if there exists an $(1+\varepsilon)$-approximation algorithm for finding an optimal balanced k-star with a given core in $O(f(n))$ time, there exists an approximation algorithm for a PROCT with ratio $(1+\varepsilon)(k+3)/(k+1)$ and time complexity $O(n^k f(n) + n^3)$.

5.2.3 Approximating by 2-stars

The core of a 2-star is an edge and there are $O(n^2)$ possible cores. We now show an algorithm for an optimal 2-star of a metric graph. We define a notation for 2-stars as follows:

DEFINITION 5.6 Let (X, Y) be a partition of V, and $x \in X$ and $y \in Y$. We use $\text{Tstar}(x, y, X, Y)$ to denote the 2-star whose edge set is

$$\{(x, v) | v \in X, v \neq x\} \cup \{(y, v) | v \in Y, v \neq y\} \cup \{(x, y)\}.$$

The next result follows immediately from Lemma 5.1. Recall that $R = r(T)$ is the total vertex weight.

FACT 5.1
Let $T = \text{Tstar}(x, y, X, Y)$.
$$C_p(T) = 2r(X)r(Y)w(x, y)$$
$$+ 2\sum_{v \in X} r(v)(R - r(v))w(x, v) + 2\sum_{v \in Y} r(v)(R - r(v))w(y, v).$$

5.2.3.1 Algorithm

Some readers may wonder if the minimum p.r.c. cost 2-star (or even k-star) can be found by the algorithm for its unweighted counterpart in Chapter 4. Unfortunately the answer is "No."

Any k-star can be described by a triple $(S, \tau, \mathcal{L})$, where $S = \{v_1, \ldots, v_k\} \subseteq V$ is the set of k distinguished vertices which may have degree more than one, τ is a spanning tree topology on S, and $\mathcal{L} = (L_1, \ldots, L_k)$, where $L_i \subseteq V - S$ is the set of vertices connected to vertex $v_i \in S$. Let $A = (n_1, \ldots, n_k)$ be a nonnegative k-vector (a vector whose components are k nonnegative integers) such that $\sum_{i=1}^{k} n_i = n - k$. We say that a k-star $(S, \tau, \mathcal{L})$ has the configuration (S, τ, A) if $n_i = |L_i|$ for all $1 \leq i \leq k$. For a fixed k, the total number of configurations is $O(n^{2k-1})$ since there are $\binom{n}{k}$ choices for S, k^{k-2} possible tree topologies on k vertices, and $\binom{n-1}{k-1}$ possible such k-vectors. Note that any two k-stars with the same configuration have the same routing load on their corresponding edges.

Any vertex v in $V - S$ that is connected to a node $s \in S$ contributes to the (standard) routing cost of a term of $w(v, s)$ multiplied by its routing load of $2(n-1)$. Since all these routing loads are the same, the best way of connecting the vertices in $V - S$ to nodes in S can be obtained by solving a minimum-cost perfect matching problem.

However, we cannot find the minimum p.r.c. cost k-star by the above method because we do not know how to find (in polynomial time) the best way to connect the vertices in $V - S$ to nodes in S even for a fixed configuration. Two k-stars with the same configuration may have *different p.r.c. loads* on their corresponding edges. If we modify the definition of the configuration so that $n_i = r(L_i)$, then it may be possible to find the best leaf connection for a fixed configuration in polynomial time. But the number of configurations might be exponential.

Another question is whether a minimum p.r.c. cost k-star can be found by an incremental method similar to the one in Chapter 4. Let us focus on the case $k = 2$. For fixed x and y, let X_i and Y_i be the vertex sets such that $\text{Tstar}(x, y, X_i, Y_i)$ is a minimum routing cost 2-star with exactly i leaves connected to x for $i = 0, 1, \ldots, n-2$. The key point of the incremental method is the following property:

> There always exists a vertex $v \in Y_i$ such that $X_{i+1} = X_i \cup \{v\}$. Therefore, instead of solving many assignment problems, all X_i can be found one by one.

However, the property does not hold for a p.r.c. cost 2-star. For example, assume $X_1 = \{v_1\}$ and $Y_1 = \{v_2, v_3\}$ (Figure 5.3). All vertex weights on x,y,v_1 are small, say 1, and $r(v_2) = r(v_3) = a$ is a large number. The vertex weights are set in such a way that the p.r.c. load on edge (x, y) will be very large if $\{v_1, v_2\}$ or $\{v_1, v_3\}$ is the set of leaves connected to x. The large load will force $X_2 = \{v_2, v_3\}$, and this is a counterexample of the above property.

Optimal Communication Spanning Trees

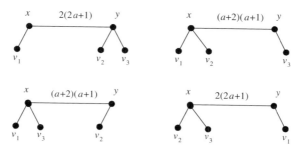

FIGURE 5.3: A counterexample of the incremental method.

Now let us turn to the algorithm for a minimum p.r.c. cost 2-star. To find the best partition for a specified pair of vertices x and y, we construct an auxiliary graph $H_{x,y}$, which is an undirected complete graph with vertex set V and edge length function h. The edge length h is defined as follows:

- $h(x,y) = 2r(x)r(y)w(x,y)$.

- $h(x,v) = 2r(v)(R - r(v))w(y,v) + 2r(v)r(x)w(x,y)$ for $v \notin \{x,y\}$.

- $h(y,v) = 2r(v)(R - r(v))w(x,v) + 2r(v)r(y)w(x,y)$ for $v \notin \{x,y\}$.

- $h(u,v) = 2r(u)r(v)w(x,y)$ for all $u, v \notin \{x,y\}$.

Let V_1 and V_2 be two subsets of V. We say that (V_1, V_2) is an x-y cut of $H_{x,y}$ if (V_1, V_2) forms a partition of V and $x \in V_1$ and $y \in V_2$. The cost of an x-y cut (V_1, V_2) is defined to be

$$h(V_1, V_2) = \sum_{u \in V_1, v \in V_2} h(u,v).$$

The following lemma comes directly from the above construction. Note that the 2-star is defined on the metric graph G and the cost of the cut is defined on the auxiliary graph $H_{x,y}$.

LEMMA 5.5

If (V_1, V_2) is an x-y cut of graph $H_{x,y}$, then $h(V_1, V_2) = C_p(\text{Tstar}(x, y, V_1, V_2))$.

PROOF

$$h(V_1, V_2)$$
$$= \sum_{u \in V_1, v \in V_2} h(u,v)$$

$$= \sum_{v \in V_2 - \{y\}} h(x,v) + \sum_{u \in V_1 - \{x\}} h(u,y) + \left(\sum_{\substack{u \in V_1 - \{x\} \\ v \in V_2 - \{y\}}} h(u,v) \right) + h(x,y)$$

$$= \sum_{v \in V_2 - \{y\}} (2r(v)(R - r(v))w(y,v) + 2r(v)r(x)w(x,y))$$
$$+ \sum_{u \in V_1 - \{x\}} (2r(u)(R - r(u))w(x,u) + 2r(u)r(y)w(x,y))$$
$$+ \left(\sum_{\substack{u \in V_1 - \{x\} \\ v \in V_2 - \{y\}}} 2r(u)r(v)w(x,y) \right) + 2r(x)r(y)w(x,y)$$

$$= \sum_{v \in V_2 - \{y\}} 2r(v)(R - r(v))w(y,v) + \sum_{v \in V_1 - \{x\}} 2r(v)(R - r(v))w(x,v)$$
$$+ 2r(V_1)r(V_2)w(x,y)$$
$$= C_p(\text{Tstar}(x, y, V_1, V_2)).$$

□

The above lemma implies that a minimum p.r.c. cost 2-star can be found by solving the minimum cut problems on the auxiliary graphs. Since the minimum cut of a graph can be found in $O(n^3)$, by Lemma 5.4 we have the following result:

LEMMA 5.6

A minimum p.r.c. cost 2-star can be found in $O(n^5)$ time.

Also by Lemma 5.4, a minimum p.r.c. cost 2-star is a (5/3)-approximation of a PROCT. Here we shall give a more precise analysis to show that the approximation ratio is 1.577.

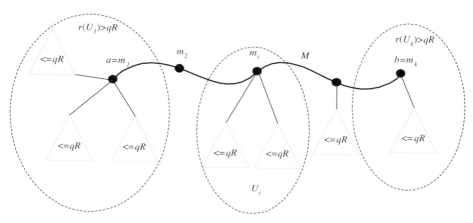

FIGURE 5.4: The approximation ratio of a 2-star.

5.2.3.2 Approximation ratio

Let $G = (V, E, w)$ and r be the input metric graph and the vertex weight of a $\triangle$PROCT problem, respectively. Also let T be an optimal spanning tree of the $\triangle$PROCT problem and m be the r-centroid of T. We are going to construct two 2-stars and show that one of them is a 1.577-approximation of T. First we establish a lower bound of the optimal cost.

Root T at its r-centroid m and let $1/3 < q < 0.5$ be a real number to be determined later. For a rooted tree T and a vertex $v \in V(T)$, we use T_v to denote the subtree rooted at v. Consider all possible vertices x such that $r(T_x) \geq qR$ and $r(T_u) < qR$ for any $u \in V(T_x) - \{x\}$. By the definition of the r-centroid, there are three cases:

1. there are two such vertices a and b;
2. there is only one such vertex $a \neq m$;
3. m is the only one such vertex.

For each case, we select two vertices. For the first case, a and b are selected. For the second case, a and m are selected, and the third case can be thought of as a special case in which the two vertices are both m. Without loss of generality, assume the two vertices be a and b, and $M = SP_T(a,b) = (a = m_1, m_2, \ldots, m_k = b)$ be the path on T. Also let U_i be the set of vertices which are connected to M at m_i for $i = 1, 2, \ldots, k$. The notations are illustrated in Figure 5.4.

LEMMA 5.7

$C_p(T) \geq 2(1-q) R \sum_x r(x) d_T(x, M) + 2q(1-q) R^2 w(M)$.

PROOF For any vertex x, let $SB(x) = \{u | SP_T(x, u) \cap M = \emptyset\}$ which is the set of vertices in the same branch of x. Note that $r(SB(x)) < qR$ by the construction of M. For vertex $x \in U_i$ and vertex $y \in U_j$, define $g(x, y) = d_T(m_i, m_j)$. Then,

$$C_p(T) = \sum_x \sum_y r(x) r(y) d_T(x, y)$$

$$\geq \sum_x \sum_{y \notin SB(x)} r(x) r(y) d_T(x, y)$$

$$= \sum_x \sum_{y \notin SB(x)} r(x) r(y) \{d_T(x, M) + d_T(y, M) + g(x, y)\}$$

$$= 2 \sum_x \sum_{y \notin SB(x)} r(x) r(y) d_T(x, M) + \sum_x \sum_{y \notin SB(x)} r(x) r(y) g(x, y)$$

$$\geq 2(1-q) R \sum_x r(x) d_T(x, M) + \sum_x \sum_{y \notin SB(x)} r(x) r(y) g(x, y).$$

Without loss of generality, we assume $r(U_1) \geq r(U_k)$. For the second term,

$$\sum_x \sum_{y \notin SB(x)} r(x) r(y) g(x, y)$$

$$= 2 \sum_{i<j} r(U_i) r(U_j) d_T(m_i, m_j)$$

$$\geq 2 r(U_1) r(U_k) d_T(m_1, m_k)$$

$$+ 2 \sum_{i=2}^{k-1} r(U_i) \left(r(U_1) d_T(m_1, m_i) + r(U_k) d_T(m_k, m_i) \right)$$

$$\geq 2 r(U_1) r(U_k) w(M) + 2 \sum_{i=2}^{k-1} r(U_i) r(U_k) w(M)$$

$$= 2 w(M) r(U_k) (R - r(U_k)).$$

By the construction of M and $q < 0.5$, we have $r(U_1) \geq r(U_k) \geq qR$, and then $qR \leq r(U_k) \leq (1-q) R$. Thus $r(U_k) (R - r(U_k)) \geq q(1-q) R^2$ and this completes the proof. □

Construct two 2-stars

$$T^* = \text{Tstar}(a, b, V - U_k, U_k),$$

and

$$T^{**} = \text{Tstar}(a, b, U_1, V - U_1).$$

We claim that one of the two 2-stars is an approximation solution with approximation ratio $\max\{\frac{1}{1-q}, (1 - 2q^2)/(2q(1-q))\}$. First, we show the following lemma:

LEMMA 5.8
$C_p(T^*) + C_p(T^{**}) \leq 4R \sum_{v \in V} r(v) d_T(v, M) + 2(1 - 2q^2) R^2 w(M)$.

PROOF By Fact 5.1,

$$C_p(T^*) = 2 \sum_{v \notin U_k} r(v)(R - r(v)) w(v, a) + 2 \sum_{v \in U_k} r(v)(R - r(v)) w(v, b)$$
$$+ 2r(U_k)(R - r(U_k)) w(a, b)$$
$$\leq 2R \sum_{v \notin U_k} r(v) w(v, a) + 2R \sum_{v \in U_k} r(v) w(v, b)$$
$$+ 2r(U_k)(R - r(U_k)) w(a, b).$$

Similarly,

$$C_p(T^{**}) \leq 2R \sum_{v \in U_1} r(v) w(v, a) + 2R \sum_{v \notin U_1} r(v) w(v, b)$$
$$+ 2r(U_1)(R - r(U_1)) w(a, b).$$

By the triangle inequality, $w(x, y) \leq d_T(x, y)$ for any vertices x and y. Therefore, for any vertex $v \in U_1$, $w(v, a) \leq d_T(v, M)$. Similarly, $w(v, b) \leq d_T(v, M)$ for any vertex $v \in U_k$. For any vertex $v \notin U_1 \cup U_k$, by the triangle inequality, $w(v, a) + w(v, b) \leq 2 d_T(v, M) + w(M)$. We have

$$C_p(T^*) + C_p(T^{**})$$
$$\leq 4R \sum_{v \in V} r(v) d_T(v, M) + 2R \sum_{v \notin U_1 \cup U_k} r(v) w(M)$$
$$+ 2r(U_k)(R - r(U_k)) w(a, b) + 2r(U_1)(R - r(U_1)) w(a, b)$$
$$= 4R \sum_{v \in V} r(v) d_T(v, M) + 2(R^2 - r(U_1)^2 - r(U_k)^2) w(M)$$
$$\leq 4R \sum_{v \in V} r(v) d_T(v, M) + 2(1 - 2q^2) R^2 w(M).$$

□

LEMMA 5.9
There is a 2-star which is a 1.577-approximation solution of the $\triangle PROCT$ problem.

PROOF Trivially, T^* and T^{**} are both 2-stars. By Lemma 5.8, we have

$$\min\{C_p(T^*), C_p(T^{**})\} \leq 2R \sum_{v \in V} r(v) d_T(v, M) + (1 - 2q^2) R^2 w(M).$$

By Lemma 5.7, the approximation ratio is

$$\max\left\{\frac{1}{1-q}, \frac{1-2q^2}{2q(1-q)}\right\},$$

in which $1/3 < q < 1/2$. By setting $q = \frac{\sqrt{3}-1}{2} \simeq 0.366$, we get the ratio 1.577.
□

Combining Lemma 5.6 and Lemma 5.9, we have the next result.

THEOREM 5.1
A PROCT of a general graph can be approximated with ratio 1.577 in $O(n^5)$ time.

It can be easily verified that the analysis of approximation ratio in this section can also be applied to the MRCT problem. Together with the time complexity of finding a minimum routing cost 2-star in Chapter 4, we have the following corollary.

COROLLARY 5.4
An MRCT of a general graph can be approximated with ratio 1.577 in $O(n^4)$ time.

5.2.4 A polynomial time approximation scheme

We now show that the PROCT problem admits a PTAS. Instead of finding a minimum p.r.c. cost k-star exactly, the PTAS finds an approximation of an optimal balanced k-star. By Corollary 5.3, if there exists a PTAS for an optimal balanced k-star with a given core, there also exists a PTAS for a PROCT.

5.2.4.1 A pseudo-polynomial time algorithm

We start at a pseudo-polynomial time algorithm. Let $G = (V, E, w)$ be a metric graph and A be a tree with $|V(A)| = k$ as the core. Let $U = V - V(A)$. The goal is to connect every vertex in U to the core so as to make the p.r.c. cost as small as possible.

For each vertex $v \in U$, we regard v as a super node consisting of $r(v)$ nodes of weight one and connected by zero-length edges (Figure 5.5). Since all these nodes have weight one, by the technique in Section 4.5.5, the best leaf connection can be found by solving a series of assignment problems. The time complexity is $O(R^k)$, in which $R = r(V)$ is the total vertex weight. We state the result in the next lemma.

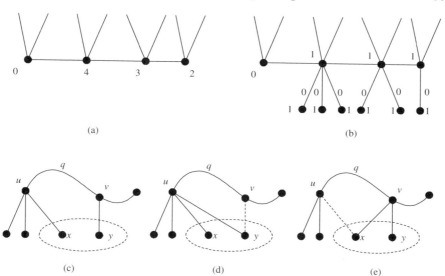

FIGURE 5.5: (a) Some vertices with nonnegative integer weights; (b) transformation to 0-1 weight vertices; (c) two vertices x and y from the same super node are connected to different nodes of the core; (d) re-connect y to u; (e) re-connect x to v.

LEMMA 5.10
For a metric graph $G = (V, E, w)$ with vertex weight r, an optimal balanced k-star with a given core can be found in $O(r(V)^k)$ time.

To show the correctness of the above lemma, we need another property.

FACT 5.2
In the best leaf connection, two nodes from the same super node are never connected to different nodes of the core.

We show the fact by contradiction. As in Figure 5.5(c), let x and y be two vertices from the same super node, and u and v be two vertices of the core. Suppose that, in an optimal tree T, x and y are connected to u and v respectively. We shall show that there is another tree whose cost is less than that of T. Let T_1 be the tree obtained by re-connecting y to u and T_2 be the tree obtained by re-connecting x to v, as frames (d) and (e) in Figure 5.5. Let's consider the cost of T_1.

$$C_p(T_1, r) = C_p(T, r) - 2\sum_{i \in V} r(i) d_T(y, i) + 2\sum_{i \in V} r(i) d_{T_1}(y, i).$$

Since
$$\sum_{i\in V} r(i)d_T(y,i) = \sum_{i\neq y} r(i)\left(w(y,v) + d_T(v,i)\right)$$
$$= (R-1)w(y,v) + \sum_{i\neq y} r(i)d_T(v,i)$$

and
$$\sum_{i\in V} r(i)d_{T_1}(y,i) = (R-1)w(y,u) + \sum_{i\neq y} r(i)d_{T_1}(u,i),$$

we have
$$C_p(T_1,r) = C_p(T,r) - 2\left((R-1)w(y,v) + \sum_{i\neq y} r(i)d_T(v,i)\right)$$
$$+ 2\left((R-1)w(y,u) + \sum_{i\neq y} r(i)d_{T_1}(u,i)\right). \tag{5.1}$$

Similarly,
$$C_p(T_2,r) = C_p(T,r) - 2\left((R-1)w(x,u) + \sum_{i\neq x} r(i)d_T(u,i)\right)$$
$$+ 2\left((R-1)w(x,v) + \sum_{i\neq x} r(i)d_{T_2}(v,i)\right). \tag{5.2}$$

Recall that $w(x,a) = w(y,a)$ for any vertex a of the core. Summing up (5.1) and (5.2), we obtain

$$\frac{1}{2}(C_p(T_1,r) + C_p(T_2,r))$$
$$= C_p(T,r) + \left(\sum_{i\neq y} r(i)d_{T_1}(u,i) - \sum_{i\neq x} r(i)d_T(u,i)\right)$$
$$+ \left(\sum_{i\neq x} r(i)d_{T_2}(v,i) - \sum_{i\neq y} r(i)d_T(v,i)\right)$$
$$= C_p(T,r) + (d_{T_1}(u,x) - d_T(u,y)) + (d_{T_2}(v,y) - d_T(v,x))$$
$$= C_p(Y,r) + (w(u,x) - d_T(u,v) - w(u,y)) + (w(v,y) - d_T(v,u) - w(u,x))$$
$$= C_p(T,r) - 2d_T(u,v).$$

Since it is assumed that there is no zero-length edge, we have
$$\min\{C_p(T_1,r), C_p(T_2,r)\} < C_p(T,r).$$
This is a contradiction to the assumption that the cost of T is minimum.

5.2.4.2 A scaling and rounding algorithm

The main drawback of the pseudo-polynomial time algorithm is that the time complexity depends on the total weight of the vertices. To reduce the time complexity, a natural idea is to scale down the weight of each vertex by a common factor. This is the method of the PTAS in this section. Since the algorithm works only on vertices of integer weights, there are rounding errors. To ensure the quality of the solution, some details of the algorithm should be designed carefully and we need to show that the rounding errors on the vertex weights do not affect too much the cost of the solution.

DEFINITION 5.7 *Let r be a vertex weight of a graph G and q a positive number. We use $q \cdot r$ to denote the vertex weight function defined by $(q \cdot r)(v) = q \times r(v)$ for every $v \in V$.*

Let T be any spanning tree of a graph and r_1, r_2 be vertex weight functions. By definitions, it is easy to see that $c(T, q \cdot r) = q^2 c(T, r)$ for any nonnegative number q. Also, if $r_1(v) \leq r_2(v)$ for any vertex v, then $c(T, r_1) \leq c(T, r_2)$.

By a selected threshold, we first divide $U = V(G) - V(A)$ into a light part and a heavy part according to their weights. Then, for each vertex in the heavy part, we scale down their weights by a scaling factor and round them to integers. When the weights are all integers, the number of configurations is polynomial in the total scaled weight of the vertices in the heavy parts. Therefore, the best connection (with respect to the scaled weights) can be determined by a pseudo-polynomial time algorithm. Finally, each vertex in the light part is connected to its closest vertex in A. It will be shown that the approximation ratio and time complexity are determined by k, the scaling factor, and the threshold for dividing the vertices into the light and heavy parts. The PTAS is given below.

Algorithm: PTAS_STAR
Input: A metric graph $G = (V, E, w)$ with vertex weight r, a tree A in G with $|V(A)| = k$, a positive number $\lambda < 1$ and a positive integer q.
Output: A k-star with core A.
/* assume $V(A) = \{a_i | 1 \leq i \leq k\}$ and $U = \{1..n - k\}$
in which $r(i) \leq r(i+1)$ for each $i \in U$.
1: Find the maximum j such that $r(\{1..j\}) \leq \lambda R$.
Let $V_L = \{1..j\}$ and $V_H = \{j+1..n-k\}$ and $\mu = r(j+1)$.
2: Let $\bar{r}(v) = \lfloor qr(v)/\mu \rfloor$ for each $v \in V_H$;
and $\bar{r}(v) = qr(v)/\mu$ for each $v \in V(A)$.
3: Find an optimal k-star T_1 with respect to $\bar{r}$.
4: Construct T_A from T_1 by connecting each vertex in V_L to the closest vertex in A.
5: Output T_A.

By Lemma 5.10, it takes $O((qR/\mu)^k)$ time for finding the best connection for vertices in V_H. From the choice of μ, $(j+1)\mu \geq \sum_{1 \leq i \leq j+1} r(i) > \lambda R$. We have $\mu > \lambda R/n$, and the time complexity is stated in the next lemma.

LEMMA 5.11
The time complexity of Algorithm PTAS_STAR *is* $O((qn/\lambda)^k)$.

5.2.4.3 Approximation ratio

In the algorithm, T_A is obtained by inserting light leaves into T_1. Since they are vertices of small weight, the load on each core edge will not be increased too much. Consider the light leaves one by one, as they are inserted. For any $e \in E(A)$ and any $v \in V_L$, whenever v is inserted, the load increase on e is no more than $2r(v)R$. Summing over all $v \in V_L$, the total load increase on e is no more than $2RR_L$, where $R_L = r(V_L)$. We state it as the following lemma.

LEMMA 5.12
For each edge $e \in E(A)$, $(l(T_A, r, e) - l(T_1, r, e)) \leq 2RR_L$.

Now we show the approximation ratio below:

LEMMA 5.13
Let X_A *be an optimal balanced k-star with core* A. *Whenever* X_A *exists,*
$c(T_A, r) \leq ((1+q^{-1})^2 + \lambda(k+3)^2/(k+1))\, c(X_A, r)$.

PROOF Let X_1 be the tree obtained by deleting the leaf set V_L from X_A. Since $C_p(T_1, \bar{r})$ is minimum, we have

$$C_p(T_1, \bar{r}) \leq C_p(X_1, \bar{r}). \tag{5.3}$$

Since $\bar{r}(v) \leq qr(v)/\mu$ for any $v \in V(A) \cup V_H$,

$$C_p(X_1, \bar{r}) \leq C_p\left(X_1, \left(\frac{q}{\mu}\right) \cdot r\right) = \left(\frac{q}{\mu}\right)^2 C_p(X_1, r). \tag{5.4}$$

By (5.3) and (5.4), we obtain

$$C_p(T_1, \bar{r}) \leq \left(\frac{q}{\mu}\right)^2 C_p(X_1, r). \tag{5.5}$$

For $v \in V_H$, $qr(v)/\mu \leq \bar{r}(v) + 1 \leq (1+q^{-1})\bar{r}(v)$; and for $v \in V(A)$, $qr(v)/\mu = \bar{r}(v) \leq (1+q^{-1})\bar{r}(v)$. Therefore, $qr(v)/\mu \leq (1+q^{-1})\bar{r}(v)$ for any $v \in V(A) \cup V_H$. Then

$$C_p\left(T_1, \frac{q}{\mu} \cdot r\right) \leq C_p(T_1, (1+q^{-1}) \cdot \bar{r}) = (1+q^{-1})^2 C_p(T_1, \bar{r}). \tag{5.6}$$

Since $C_p(T_1, r) = (\mu/q)^2 C_p(T_1, (q/\mu) \cdot r)$, by (5.5) and (5.6),

$$C_p(T_1, r) \leq \left(\frac{\mu}{q}\right)^2 (1 + q^{-1})^2 C_p(T_1, \bar{r}) \leq (1 + q^{-1})^2 C_p(X_1, r). \quad (5.7)$$

For a subset B of leaves in a tree T, let $C_L(T, B)$ denote the total p.r.c. routing cost on the edges incident with the leaves in B, i.e., $C_L(T, B) = 2\sum_{i \in B} r(i)(r(T) - r(i)) w(i, f(T, i))$, where $f(T, i)$ is the vertex adjacent to leaf i in T. By Lemma 5.1,

$$C_p(T_A, r) = \sum_{e \in E(A)} l(T_A, r, e) w(e) + C_L(T_A, V_H) + C_L(T_A, V_L).$$

Since

$$C_L(T_A, V_H) - C_L(T_1, V_H) = 2 \sum_{i \in V_H} r(i) R_L w(i, f(T_A, i))$$

$$\leq 2 \sum_{i \in V_H} r(i) R_L \left(w(i, f(X_A, i)) + w(A)\right)$$

$$= C_L(X_A, V_H) - C_L(X_1, V_H) + 2 R_H R_L w(A),$$

we have

$$C_p(T_A, r) \leq C_p(T_1, r) + \sum_{e \in E(A)} (l(T_A, r, e) - l(T_1, r, e)) w(e) + C_L(T_A, V_L)$$

$$+ C_L(X_A, V_H) - C_L(X_1, V_H) + 2 R_H R_L w(A). \quad (5.8)$$

Similarly,

$$C_p(X_A, r) \geq C_p(X_1, r) + C_L(X_A, V_H) - C_L(X_1, V_H) + C_L(X_A, V_L). \quad (5.9)$$

Since the light leaves on T_A are connected to the closest nodes in $V(A)$, $C_L(T_A, V_L) \leq C_L(X_A, V_L)$. Then by (5.7), (5.8), (5.9), and Lemma 5.12, we have

$$C_p(T_A, r) \leq (1 + q^{-1})^2 C_p(X_A, r) + \sum_{e \in E(A)} (l(T_A, r, e) - l(T_1, r, e)) w(e)$$

$$+ 2 R_H R_L w(A)$$

$$\leq (1 + q^{-1})^2 C_p(X_A, r) + 2 R_L (R + R_H) w(A). \quad (5.10)$$

If $k = 1$, the lemma holds trivially since there is no edge in A. In the following, we assume $k > 1$. For a balanced k-star, since its core is a minimal $(2/(k+3))$-separator, the routing load on any core edge is no less than $2(2R/(k+3))(R - 2R/(k+3)) = 4R^2(k+1)(k+3)^{-2}$. We have $l(X_A, r, e) \geq 4R^2(k+1)(k+3)^{-2}$ for each $e \in E(A)$. Thus, $C_p(X_A, r) \geq 4R^2(k+1)(k+3)^{-2} w(A)$. Since $R_L \leq \lambda R$, by (5.10) we have

$$C_p(T_A, r) \leq \left((1 + q^{-1})^2 + \lambda(2 - \lambda)(k+3)^2/(2k+2)\right) C_p(X_A, r)$$

$$\leq \left((1 + q^{-1})^2 + \lambda(k+3)^2/(k+1)\right) C_p(X_A, r).$$

THEOREM 5.2
Algorithm PTAS_STAR *is a PTAS for an optimal balanced k-star with a given core. For any positive integer q and positive number $\lambda < 1$, the time complexity is $O((nq/\lambda)^k)$ and approximation ratio is $((1+q^{-1})^2 + \lambda(k+3)^2/(k+1))$.*

PROOF The time complexity and approximation ratio are shown in Lemmas 5.11 and 5.13 respectively. It can be easily shown that the approximation ratio approaches to 1 as q and λ^{-1} go to infinity. Therefore, for any desired approximation ratio $1 + \varepsilon > 1$, we can choose suitable q and λ, and the time complexity is polynomial when they are fixed. □

Together with Corollary 5.3, we conclude the section by the next theorem.

THEOREM 5.3
The PROCT problem on general graphs admits a PTAS.

5.3 Sum-Requirement

The formal definition of the problem is in the following.

> PROBLEM: Optimal Sum-Requirement Communication Spanning Trees (SROCT)
> INSTANCE: A $G = (V, E, w)$ with vertex weight $r : V \to Z_0^+$.
> GOAL: Find a spanning tree T of minimum s.r.c. cost.

Recall that the s.r.c. routing cost of a tree T is defined by $C_s(T) = \sum_{u,v}(r(u) + r(v))d_T(u,v)$. Similar to the PROCT problem, the SROCT problem includes the MRCT problem as a special case and is therefore NP-hard. The s.r.c. cost of a tree can also be computed by summing the routing costs of edges. The only difference is the definition of routing load.

DEFINITION 5.8 *Let T be any spanning tree of a graph G, and r a vertex weight function. For any edge $e = (u,v) \in E(T)$, we define the s.r.c. routing load on the edge e to be $l_s(T,r,e) = 2(r(T_u) + r(T_v))$, where T_u and T_v are the two subgraphs obtained by removing e from T. The s.r.c. routing cost on the edge e is defined to be $l_s(T,r,e)w(e)$.*

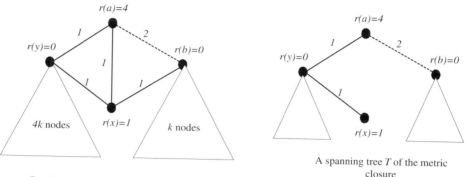

Graph G. (a,b) is not an edge in G

A spanning tree T of the metric closure

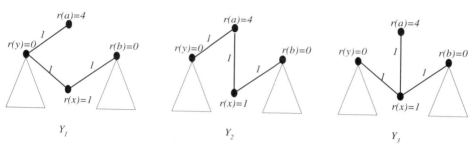

FIGURE 5.6: A tree with bad edges may have less s.r.c. cost. The triangles represent nodes of zero weight and connected by zero-length edges.

LEMMA 5.14

Let T be any spanning tree of a graph $G = (V, E, w)$ and r be a vertex weight function. $C_s(T) = \sum_{e \in E(T)} l_s(T, r, e) w(e)$.

In this section, we focus on the approximation algorithm for an SROCT. For the PROCT problem, it has been shown that an optimal solution for a graph has the same value as the one for its metric closure. In other words, using bad edges cannot lead to a better solution. However, the SROCT problem has no such a property. For example, consider the graph G in Figure 5.6. The edge (a, b) is not in $E(G)$, and T is a spanning tree of the metric closure of G. All three possible spanning trees of G are Y_1, Y_2 and Y_3. It will be shown that the s.r.c cost of T is less than that of Y_i for $i = 1, 2, 3$.

To compare the s.r.c costs, we can only focus on the coefficient of k in the cost. Note that only vertices a and x have nonzero weights. By Lemma 5.14, the s.r.c. cost of T can be computed as follows:

$$C_s(T)$$
$$= l_s(T, r, (a, b))w(a, b) + l_s(T, r, (a, y))w(a, y) + l_s(T, r, (y, x))w(x, y)$$
$$= 2(k(4+1) + 0(4k))2 + 2(k \times 1 + 4 \times 4k)(1) + 2(5k \times 1 + 4 \times 1)(1)$$
$$= 64k + \ldots$$

Similarly we have $C_s(Y_1) = 66k$, $C_s(Y_2) = 66k$, and $C_s(Y_3) = 90k$. The example illustrates that it is impossible to transform any spanning tree of $\bar{G}$ to a spanning tree of G without increasing the s.r.c cost for some graph G, where $\bar{G}$ is the metric closure of G. But it should be noted that the example does not disprove the possibility of reducing the SROCT problem on general graphs to its metric version.

We shall present a 2-approximation algorithm for the SROCT problem on general graphs. For each vertex v of the input graph, the algorithm finds the shortest-paths tree rooted at v. Then it outputs the shortest-paths tree with minimum s.r.c. cost. We shall show that there always exists a vertex x such that any shortest-paths tree rooted at x is a 2-approximation solution.

In the following, graph $G = (V, E, w)$ and vertex weight r is the input of the SROCT problem. We assume that $|V| = n$, $|E| = m$ and $r(V) = R$.

LEMMA 5.15
Let T be a spanning tree of G. For any vertex $x \in V$,

$$C_s(T) \leq 2 \sum_{v \in V} (nr(v) + R) \, d_T(v, x).$$

PROOF

$$C_s(T) = \sum_{u,v \in V} (r(u) + r(v)) \, d_T(u, v)$$
$$\leq \sum_{u,v \in V} (r(u) + r(v)) \, (d_T(u, x) + d_T(x, v))$$
$$= 2 \sum_{u,v \in V} (r(u) + r(v)) \, d_T(u, x)$$
$$\leq 2 \sum_{v \in V} (nr(v) + R) \, d_T(v, x).$$

☐

In the following, we use T to denote an optimal spanning tree of the SROCT problem, and use x_1 and x_2 to denote a centroid and an r-centroid of T

respectively. Let $P = SP_T(x_1, x_2)$ be the path between the two vertices on the tree. If x_1 and x_2 are the same vertex, P contains only one vertex.

LEMMA 5.16

For any edge $e \in E(P)$, the s.r.c load $l_s(T, r, e) \geq nR$.

PROOF Let T_1 and T_2 be the two subtrees resulting by deleting e from T. Assume that $x_1 \in V(T_1)$ and $x_2 \in V(T_2)$. By the definitions of centroid and r-centroid, $|V(T_1)| \geq n/2$ and $r(T_2) \geq R/2$. Then,

$$\begin{aligned} l_s(T,r,e)/2 &= |V(T_1)|r(T_2) + |V(T_2)|r(T_1) \\ &= |V(T_1)|r(T_2) + (n - |V(T_1)|)(R - r(T_2)) \\ &= 2(|V(T_1)| - n/2)(r(T_2) - R/2) + nR/2 \geq nR/2. \end{aligned}$$

$\Box$

The next lemma establishes a lower bound of the minimum s.r.c. cost. Remember that $d_T(v, P)$ denotes the shortest path length from vertex v to path P.

LEMMA 5.17

$C_s(T) \geq \sum_{v \in V} (nr(v) + R) d_T(v, P) + nRw(P)$.

PROOF For any vertex u, we define $SB(u)$ to be the set of vertices in the same branch of u. Note that $|SB(u)| \leq n/2$ and $r(SB(u)) \leq R/2$ for any vertex u by the definitions of centroid and r-centroid.

$$\begin{aligned} C_s(T) &= \sum_{u,v \in V} (r(u) + r(v)) d_T(u,v) \\ &= 2 \sum_{u,v \in V} r(u) d_T(u,v) \\ &\geq 2 \sum_{u \in V} \sum_{v \notin SB(u)} r(u) (d_T(u,P) + d_T(v,P)) \\ &\quad + 2 \sum_{u,v \in V} r(u) w(SP_T(u,v) \cap P). \end{aligned} \quad (5.11)$$

For the first term in (5.11),

$$2\sum_{u\in V}\sum_{v\notin SB(u)} r(u)\left(d_T(u,P)+d_T(v,P)\right)$$

$$=2\sum_{u\in V}\sum_{v\notin SB(u)} r(u)d_T(u,P) + 2\sum_{u\in V}\sum_{v\notin SB(u)} r(u)d_T(v,P)$$

$$\geq \sum_{u\in V} nr(u)d_T(u,P) + 2\sum_{v\in V}\sum_{u\notin SB(v)} r(u)d_T(v,P)$$

$$\geq \sum_{u\in V} nr(u)d_T(u,P) + \sum_{v\in V} Rd_T(v,P)$$

$$= \sum_{v\in V} (nr(v)+R)\,d_T(v,P). \qquad (5.12)$$

For the second term in (5.11),

$$2\sum_{u,v\in V} r(u)w(SP_T(u,v)\cap P)$$

$$= 2\sum_{u,v\in V} r(u)\left(\sum_{e\in SP_T(u,v)\cap P} w(e)\right)$$

$$= \sum_{e\in E(P)} \left(2\sum_v r(\{u|e\in E(SP_T(u,v))\})\right) w(e)$$

$$= \sum_{e\in E(P)} l_s(T,r,e)w(e)$$

$$\geq nRw(P). \qquad \text{(by Lemma 5.16)} \qquad (5.13)$$

□

The result follows (5.11), (5.12), and (5.13).

The main result of this section is stated in the next theorem.

THEOREM 5.4
There exists a 2-approximation algorithm with time complexity $O(n^2 \log n + mn)$ for the SROCT problem.

PROOF Let Y^* and Y^{**} be the shortest-path trees rooted at x_1 and x_2 respectively. Also, for any $v \in V$, let $h_1(v) = w(SP_T(v,x_1) \cap P)$ and $h_2(v) = w(SP_T(v,x_2) \cap P)$. By Lemma 5.15,

$$C_s(Y^*)/2 \leq \sum_{v\in V} (nr(v)+R)\,d_{Y^*}(v,x_1)$$

$$\leq \sum_{v\in V} (nr(v)+R)\,(d_T(v,P)+h_1(v)). \qquad (5.14)$$

Similarly

$$C_s(Y^{**})/2 \leq \sum_{v \in V} (nr(v) + R)(d_T(v, P) + h_2(v)). \qquad (5.15)$$

Since $h_1(v) + h_2(v) = w(P)$ for any vertex v, by (5.14) and (5.15), we have

$$\begin{aligned}
&\min\{C_s(Y^*), C_s(Y^{**})\} \\
&\leq (C_s(Y^*) + C_s(Y^{**}))/2 \\
&\leq \sum_{v \in V} (nr(v) + R)(2d_T(v, P) + h_1(v) + h_2(v)) \\
&= \sum_{v \in V} (nr(v) + R)(2d_T(v, P) + w(P)) \\
&= 2 \sum_{v \in V} (nr(v) + R) d_T(v, P) + 2nRw(P) \\
&\leq 2C_s(T). \qquad \text{(by Lemma 5.17)}
\end{aligned}$$

We have proved that there exists a vertex x such that any shortest-paths tree rooted at x is a 2-approximation solution. Since it takes $O(n \log n + m)$ time to construct a shortest-paths tree rooted at a given vertex and the s.r.c cost of a tree can be computed in $O(n)$ time, a 2-approximation solution of the SROCT problem can be found in $O(n^2 \log n + mn)$ time by constructing a shortest-paths tree rooted at each vertex and choosing the one with minimum s.r.c cost. □

5.4 Multiple Sources

The formal definition of the problem is in the following.

PROBLEM: p-source Minimum Routing Cost Spanning Trees (p-MRCT)
INSTANCE: A graph $G = (V, E, w)$ and a set $S \subset V$ of p sources.
GOAL: Find a spanning tree T such that the routing cost defined by $C_m(T, S) = \sum_{u \in S} \sum_{v \in V} d_T(u, v)$ is minimum.

If there is only one source, the problem is reduced to the shortest-paths tree problem, and it is always possible to find a spanning tree such that the path between the source and each vertex is a shortest path on the given graph. Therefore the 1-MRCT problem is polynomial time solvable. For the other extreme case that all vertices are sources, the problem is reduced to the minimum routing cost spanning tree (MRCT) problem, and is therefore NP-hard. Consequently the p-MRCT problem for arbitrary p is obviously NP-hard since it contains the MRCT problem as a special case. But it does not imply the NP-hardness of the p-MRCT problem for a fixed p. The readers should distinguish the difference carefully. In the p-MRCT problem for arbitrary k (or just p-MRCT problem), there is no restriction on the number of source vertices. In other words, the value p is a part of the input instance. Meanwhile, by the term "p-MRCT problem for a fixed p" we mean one of a series of p-MRCT problems, in which the value p is specified in the problem definition, such as 2-MRCT, 3-MRCT, and so on.

Recall that in the SROCT problem, the objective function is $C_s(T) = \sum_{u,v}(r(u)+r(v))d_T(u,v)$ in which $r(u)$ is the given weight of vertex u. Setting $r(v) = 1$ for each $v \in S$ and $r(v) = 0$ for other vertices, we have

$$C_s(T) = \sum_{u,v}(r(u) + r(v))d_T(u,v)$$

$$= \sum_{u \in S}\sum_{v \in S}(r(u)+r(v))d_T(u,v) + \sum_{u \in S}\sum_{v \notin S}(r(u)+r(v))d_T(u,v)$$
$$+ \sum_{u \notin S}\sum_{v \in S}(r(u)+r(v))d_T(u,v) + \sum_{u \notin S}\sum_{v \notin S}(r(u)+r(v))d_T(u,v)$$

$$= \sum_{u \in S}\sum_{v \in S} 2d_T(u,v) + \sum_{u \in S}\sum_{v \notin S} d_T(u,v) + \sum_{u \notin S}\sum_{v \in S} d_T(u,v)$$

$$= 2\left(\sum_{u \in S}\sum_{v \in S} d_T(u,v) + \sum_{u \in S}\sum_{v \notin S} d_T(u,v)\right)$$

$$= 2\sum_{u \in S}\sum_{v \in V} d_T(u,v) = 2C_m(T,S).$$

Thus the p-MRCT problem can be thought of as a special case of the SROCT problem. By Theorem 5.4, there exists a 2-approximation algorithm. We state the result of the p-MRCT problem as a theorem.

THEOREM 5.5
The p-MRCT problem is NP-hard and there exists a 2-approximation algorithm with time complexity $O(n^2 \log n + mn)$.

5.4.1 Computational complexity for fixed p

Having shown the result on p-MRCT for arbitrary p, we now turn to the problem for a fixed k. First we give some discussion on the computational complexity. We shall show that the p-MRCT problem is NP-hard even for $p = 2$ and the input is restricted to a metric graph. To show the NP-hardness for any fixed p, we define a weighted version of the 2-MRCT problem. By a transformation from the well-known *satisfiability problem*, we show the weighted 2-MRCT problem is NP-hard. Then, we show that the p-MRCT problem for

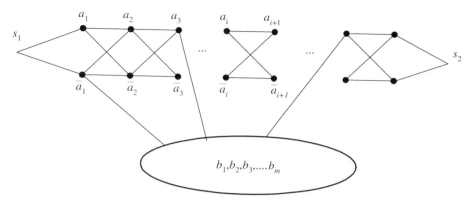

FIGURE 5.7: Transformation from SAT to 2-MRCT(α).

any fixed p can be transformed from the weighted 2-MRCT problem, and is also NP-hard.

DEFINITION 5.9 Given a set U of variables and a set X of clauses over U, the SATISFIABILITY (SAT) problem is to ask if there is a satisfying truth assignment for X.

DEFINITION 5.10 Let $G = (V, E, w)$ be a graph and $s_1, s_2 \in V$ be two given sources. For any integer $\alpha \geq 1$, the 2-MRCT(α) problem is to find a spanning tree T of G such that the weighted routing cost $C_{m2}(T, \alpha) = \sum_{v \in V}(\alpha d_T(v, s_1) + d_T(v, s_2))$ is minimum.

We now reduce the SAT problem to the 2-MRCT(α) problem. Given a set $U = \{u_1, u_2, \ldots, u_n\}$ of variables and a set $X = \{x_1, x_2, \ldots x_m\}$ of clauses as an instance of the SAT problem, we construct a graph $G = (V, E, w)$ as follows (see Figure 5.7).

- Let $A = \{a_i, \bar{a}_i | 1 \leq i \leq n\}$ and $B = \{b_i | 1 \leq i \leq m\}$. The vertex set $V = \{s_1, s_2\} \cup A \cup B$, in which s_1 and s_2 are the two sources. The vertices a_i and $\bar{a}_i$ correspond to variable u_i and negated variable $\bar{u}_i$ respectively, and b_i corresponds to clause x_i.

- The edge set E contains the following subsets.

  1. $E_1 = \{(s_1, a_1), (s_1, \bar{a}_1), (s_2, a_n), (s_2, \bar{a}_n)\}$.
  2. $E_2 = \bigcup_{1 \leq i < n}\{(a_i, a_{i+1}), (a_i, \bar{a}_{i+1}), (\bar{a}_i, a_{i+1}), (\bar{a}_i, \bar{a}_{i+1})\}$.
  3. $E_3 = \{(a_i, b_j) | u_i \in x_j, \forall i, j\} \cup \{(\bar{a}_i, b_j) | \bar{u}_i \in x_j, \forall i, j\}$.

- For any $e \in E_1 \cup E_2$, $w(e) = 1$. For any (a_i, b_j) or $(\bar{a}_i, b_j)$ in E_3, the edge weight is $L - \frac{\alpha-1}{\alpha+1}i$, in which $L = (\alpha+1)mn$.

We show that the SAT problem has a satisfying truth assignment if and only if there is a spanning tree T of G such that the weighted routing cost $c(T, \alpha) \leq \beta$, where

$$\beta = (n+1)^2\alpha + (n^2 + 4n + 1) + ((\alpha+1)L + n + 1)m.$$

PROPOSITION 5.1
If there is a truth assignment satisfying X, there exists a spanning tree Y of G such that $C_{m2}(Y, \alpha) = \beta$.

PROOF We may construct a corresponding spanning tree Y of G as follows.

- The path P_Y between the two sources has the form

$$(s_1 = v_0, v_1, v_2, \ldots, v_n, s_2),$$

in which, for $1 \leq i \leq n$, $v_i = a_i$ if u_i is assigned true and $v_i = \bar{a}_i$ otherwise.

- For any $1 \leq i \leq n$, if $v_i = a_i$, connect $\bar{a}_i$ to v_{i-1}. Otherwise, connect a_i to v_{i-1}.

- For each b_i, $1 \leq i \leq m$, insert an edge incident with both b_i and a vertex on P_Y. Such an edge always exists since at least one of the literals in x_i is assigned true.

For any $v \in \{a_i, \bar{a}_i\}$, if v is on P_Y, $d_Y(v, s_1) = i$ and $d_Y(v, s_2) = n + 1 - i$; otherwise $d_Y(v, s_1) = i$ and $d_Y(v, s_2) = n + 3 - i$. For any b_i, since it is connected to some a_j (or $\bar{a}_j$ similarly) on P_Y,

$$\alpha d_Y(b_i, s_1) + d_Y(b_i, s_2) = (\alpha+1)w(b_i, a_j) + \alpha j + (n+1-j)$$
$$= (\alpha+1)L - (\alpha-1)j + (\alpha-1)j + (n+1)$$
$$= (\alpha+1)L + n + 1.$$

Note that the cost depends only on whether it is directly connected to P_Y or not, but not on which vertex of the path it is connected to.

The routing cost of Y is given by

$$\begin{aligned}
&C_{m2}(Y,\alpha)\\
&= (\alpha+1)d_Y(s_1,s_2) + \sum_{v\in A}(\alpha d_Y(v,s_1)+d_Y(v,s_2))\\
&\quad + \sum_i(\alpha d_Y(b_i,s_1)+d_Y(b_i,s_2))\\
&= (\alpha+1)(n+1) + \alpha\sum_{i\leq n}(2i) + \sum_{i\leq n}(2n+4-2i) + m((\alpha+1)L+n+1)\\
&= (n+1)^2\alpha + (n^2+4n+1) + ((\alpha+1)L+n+1)m\\
&= \beta.
\end{aligned}$$

□

PROPOSITION 5.2
Let T be the 2-MRCT(α) of G. If $c(T,\alpha) \leq \beta$, the path P_T from s_1 to s_2 on T has the form $(s_1, v_1, v_2, \ldots, v_n, s_2)$, in which $v_i \in \{a_i, \bar{a}_i\}$ for $1 \leq i \leq n$.

PROOF If the proposition is wrong, we have that P_T contains some vertices in B or that it contains more than n vertices in A. In the following, we show that both cases lead to contradictions.

- Suppose that P_T contains some b_i. It implies that $d_T(s_1,s_2) > 2L-n$ since $d_G(b_i,s_1) > L$ and $d_G(b_i,s_2) > L-n$. For any vertex $v \in A$,

$$\alpha d_T(v,s_1) + d_T(v,s_2) \geq d_T(s_1,s_2) > 2L-n.$$

For any vertex $v \in B$,

$$\alpha d_T(v,s_1) + d_T(v,s_2) \geq \alpha d_G(v,s_1) + d_G(v,s_2) > (\alpha+1)L-n.$$

Therefore,

$$\begin{aligned}
C_{m2}(T,\alpha) &> 2n(2L-n) + m((\alpha+1)L-n)\\
&= 4nL - 2n^2 - mn + (\alpha+1)mL\\
&= 4n(\alpha+1)mn - 2n^2 - mn + (\alpha+1)mL\\
&> 4n^2m\alpha + n^2m + (\alpha+1)mL.
\end{aligned}$$

Comparing with

$$\beta = (n+1)^2\alpha + (n^2+4n+1) + ((\alpha+1)L+n+1)m,$$

we have $C_{m2}(T,\alpha) > \beta$, and it is a contradiction.

- Suppose that the path P_T contains more than n vertices in A. It implies that $d_T(s_1, s_2) > n + 1$ and there exists the smallest i such that both a_i and $\bar{a}_i$ are on the path. By the definition of G, without loss of generality, we may assume that the path $P_T = (\ldots, a_i, a_{i+1}, \bar{a}_i, \bar{a}_{i+1}, \ldots)$. However, if we replace edge $(a_{i+1}, \bar{a}_i)$ with $(a_i, \bar{a}_{i+1})$, we may obtain another spanning tree and its cost is smaller than T since the costs for a_{i+1} and $\bar{a}_i$ are decreased and the costs for any other vertex is not increased.

$\square$

PROPOSITION 5.3
Let T be the 2-MRCT(α) of G. If $C_{m2}(T, \alpha) \leq \beta$, there is a truth assignment satisfying X.

PROOF By Proposition 5.2, the path P_T from s_1 to s_2 on T has the form $(s_1, v_1, v_2, \ldots, v_n, s_2)$, in which $v_i \in \{a_i, \bar{a}_i\}$ for $1 \leq i \leq n$. For any a_i (or $\bar{a}_i$ similarly) on P_T,

$$\alpha d_T(a_i, s_1) + d_T(a_i, s_2) = i\alpha + n + 1 - i.$$

For any a_i (or $\bar{a}_i$ similarly) not on P_T,

$$\alpha d_T(a_i, s_1) + d_T(a_i, s_2)$$
$$= \min_{v \in V(P)} \{(\alpha + 1)d_T(a_i, v) + \alpha d_T(v, s_1) + d_T(v, s_2)\}$$
$$\geq (\alpha + 1) + \alpha(i - 1) + (n + 1 - (i - 1))$$
$$= i\alpha + n + 3 - i.$$

The lower bound happens when a_i is connected to v_{i-1} on P_T. For any b_i, similar to the proof of Proposition 5.1,

$$\alpha d_T(b_i, s_1) + d_T(b_i, s_2) \geq (\alpha + 1)L + n + 1.$$

The lower bound happens when b_i is connected to some vertex on P_T. Consequently,

$$C_{m2}(T, \alpha)$$
$$\geq (\alpha + 1)(n + 1) + \sum_{i=1}^{n}((i\alpha + n + 1 - i) + (i\alpha + n + 3 - i))$$
$$+ ((\alpha + 1)L + n + 1)m$$
$$= (n + 1)^2 \alpha + (n^2 + 4n + 1) + ((\alpha + 1)L + n + 1)m = \beta.$$

The lower bound happens when each vertex in B is connected to some v_i on P_T. It implies that, for each clause in X, there is a literal assigned truth, and X is satisfiable. $\square$

THEOREM 5.6
For any fixed integer $\alpha \geq 1$, the 2-MRCT(α) problem is NP-hard.

PROOF By Proposition 5.1 and 5.3, we have transformed the SAT problem to the 2-MRCT(α) problem. Given an instance of the SAT problem, in polynomial time, we can construct an instance of the 2-MRCT(α) problem. If there is a polynomial time algorithm for finding an optimal solution of the 2-MRCT(α) problem, it can also be used to solve the SAT problem. Therefore the 2-MRCT(α) problem is NP-hard since the SAT problem is NP-complete [24, 43]. □

The result of the NP-hardness can be easily extended to the metric graphs.

COROLLARY 5.5
For any fixed integer $\alpha \geq 1$, the 2-MRCT(α) problem is NP-hard even for metric graphs.

PROOF Let G be the constructed graph in Theorem 5.6, and $\overline{G} = (V, V \times V, \overline{w})$ be the metric closure of G, that is, $\overline{w}(u,v) = d_G(u,v)$ for all $u, v \in V$. Clearly $\overline{G}$ is a metric graph. The transformation is similar to Propositions 5.1–5.3. □

Let $p \geq 1$ be any fixed integer. We can easily transform the 2-MRCT(p) problem to the p-MRCT by duplicating p copies of the source s_1. The next corollary is obvious and the proof is omitted.

COROLLARY 5.6
For any fixed integer $p \geq 1$, the p-MRCT is NP-hard even for metric inputs.

Since the OCT problem includes the MRCT problem as a special case, we have the following result.

COROLLARY 5.7
For any fixed integer $p \geq 1$, the p-OCT is NP-hard even for metric inputs.

5.4.2 A PTAS for the 2-MRCT

For a tree T and $S \subset V(T)$, the routing cost of T with a source set S is defined by $C_m(T, S) = \sum_{s \in S} \sum_{v \in V(T)} d_T(s, v)$. For the 2-MRCT problem, since there are only two sources s_1 and s_2, we use $C_m(T, s_1, s_2)$ to denote the routing cost. For any $v \in V$, $d_T(v, s_1) + d_T(v, s_2) = w(P) + 2d_T(v, P)$.

Summing over all vertices in V, we obtain the next formula for computing the routing cost.

LEMMA 5.18
Let T be a spanning tree of $G = (V, E, w)$ and P be the path between s_1 and s_2 on T. $C_m(T, s_1, s_2) = n \times w(P) + 2\sum_{v \in V} d_T(v, P)$, in which $n = |V|$.

Once a path P between the two sources has been chosen, by Lemma 5.18, it is obvious that the best way to extend P to a spanning tree is to add the shortest-paths forest using the vertices of P as multiple roots, i.e., the distance from every other vertex to the path is made as small as possible. The most time-efficient algorithm for the shortest-paths tree depends on the graph. In the remaining paragraphs of this section, the time complexity for finding a shortest-paths tree/forest of a graph G is denoted by $f_{SP}(G)$. For general graphs, $f_{SP}(G)$ is $O(|E| + |V| \log |V|)$, and it is $O(|E|)$ for graphs with integer weights. However, the time complexity is $O(n^2)$ for dense graphs of which the number of edges is $\theta(n^2)$.

5.4.2.1 A simple 2-approximation algorithm

Before showing the PTAS, we give a simple 2-approximation algorithm, and then generalize the idea to the PTAS.

Algorithm: 2MRCT_1
Input: A graph $G = (V, E, w)$ and $s_1, s_2 \in V$.
Output: A spanning tree T of G.
1:    Find a shortest path P between s_1 and s_2 on G.
2:    Find the shortest-paths forest with multiple roots in $V(P)$.
3:    Output the tree T which is the union of the forest and P.

We are going to show the performance of the algorithm. First we establish a lower bound of the optimum. Let Y be an optimal tree of the 2-MRCT problem with input graph G and two sources s_1 and s_2. Since $d_Y(v, s_i) \geq d_G(v, s_i)$ for any vertex v and $i \in \{1, 2\}$,

$$C_m(Y, s_1, s_2) \geq \sum_{v \in V} (d_G(v, s_1) + d_G(v, s_2)). \tag{5.16}$$

By Lemma 5.18, we have

$$C_m(Y, s_1, s_2) \geq n d_G(s_1, s_2). \tag{5.17}$$

By (5.16) and (5.17), we have

$$C_m(Y, s_1, s_2) \geq \frac{1}{2}\sum_v (d_G(v, s_1) + d_G(v, s_2)) + \frac{n}{2} d_G(s_1, s_2). \tag{5.18}$$

Let T be the tree constructed by Algorithm 2MRCT_1 and P be the shortest between the two sources. Since each vertex is connected to P by a shortest path to any of the vertices of P at Step 2, for any vertex v,

$$d_T(v, P) \leq \min\{d_G(v, s_1), d_G(v, s_2)\}$$
$$\leq \frac{1}{2}(d_G(v, s_1) + d_G(v, s_2)).$$

By Lemma 5.18,

$$C_m(T, s_1, s_2) = n \times w(P) + 2 \sum_{v \in V} d_T(v, P)$$
$$\leq n d_G(s_1, s_2) + \sum_v (d_G(v, s_1) + d_G(v, s_2)).$$

Comparing with (5.18), we have $C_m(T, s_1, s_2) \leq 2 C_m(Y, s_1, s_2)$ and T is a 2-approximation of a 2-MRCT.

The total time complexity is dominated by the step of finding the shortest-paths tree algorithm, which is $f_{SP}(G)$. Therefore we have the next result.

LEMMA 5.19
The 2MRCT_1 *algorithm finds a 2-approximation of a 2-MRCT of a graph G in $f_{SP}(G)$ time.*

The 2-MRCT problem is a special case of the SROCT problem. By the result in the last section, a 2-MRCT can be approximated with error ratio 2 and time complexity $O(n f_{SP}(G))$. The simple algorithm ensures the same error ratio and is more efficient in time. The ratio shown in Lemma 5.19 is tight in the sense that there is an instance such that the spanning tree constructed by the algorithm has a routing cost twice as the optimum. Consider the complete graph in which $w(v, s_1) = w(v, s_2) = 1$ and $w(s_1, s_2) = 2$ for each vertex v. The distance between any other pair of vertices is zero. At Step 1, the algorithm may find edge (s_1, s_2) as the path P, and then all other vertices are connected to one of the two sources. The routing cost of the constructed tree is $4n - 4$. On an optimal tree, the path between the two sources is a two-edge path, and all other vertices are connected to the middle vertex of the path. The optimal routing cost is therefore $2n$. The increased cost is due to missing the vertex on the path. On the other hand, the existence of the vertex reduces the cost at an amount of $w(P)$ for each vertex.

The worst case instance of the simple algorithm gives us some intuitions to improve the error ratio: To reduce the error, we may try to guess some vertices of the path. Let r be a vertex of the path between the two sources on an optimal tree, and U be the set of vertices connected to the path at r. If the path P found in Step 1 of the simple algorithm includes r, the distance from any vertex in U to each of the sources will be no more than the corresponding

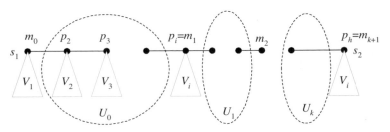

FIGURE 5.8: The definitions of the partition of the vertices.

distance on the optimal tree. In addition, the vertex r partitions the path into two subpaths. The maximal increased cost by one of the vertices is the length of the subpath instead of the whole path. The PTAS is to guess some of the vertices of the path, which partition the path such that the number of vertices connected to each subpath is small enough. We now describe the PTAS and the analysis precisely.

5.4.2.2   A PTAS

In the remaining paragraphs, let Y be a 2-MRCT of $G = (V, E, w)$ and $n = |V|$. Also let $P = (p_1 = s_1, p_2, p_3, \ldots, p_h = s_2)$ be the path between s_1 and s_2 on Y. Define V_i, $1 \leq i \leq h$, as the set of the vertices connected to P at p_i, and also $p_i \in V_i$. Let $k \geq 0$ be an integer. For $0 \leq i \leq k+1$, define $m_i = p_j$ in which j is the minimal index such that

$$\sum_{1 \leq q \leq j} |V_q| \geq \lceil i \frac{n}{k+1} \rceil.$$

By definition, $s_1 = m_0$ and $s_2 = m_{k+1}$. For $0 \leq i \leq k$, let $U_i = \bigcup_{a < j < b} V_j$ and P_i be the path from p_a to p_b, in which $p_a = m_i$ and $p_b = m_{i+1}$. Also let

$$U = V - \bigcup_{0 \leq i \leq k} U_i$$

and

$$M = \{m_i \mid 0 \leq i \leq k+1\}.$$

Note that the above definitions include the case $m_i = m_{i+1}$. In such a case, P_i contains only one vertex and U_i is empty. The definitions are shown in Figure 5.8. By the above definitions, the vertex set V is partitioned into $U, U_0, U_1, \ldots, U_k$ satisfying the properties in the following lemmas.

LEMMA 5.20
For any $v \in U$, $d_G(v, M) \leq d_Y(v, P)$.

PROOF Let $v \in V_i$. Since $p_i = m_j$ for some j,

$$d_Y(v, P) = d_Y(v, p_i) \geq d_G(v, p_i) = d_G(v, m_j) \geq d_G(v, M).$$

□

LEMMA 5.21
$\sum_{v \in U_i} d_G(v, M) \leq \sum_{v \in U_i} d_Y(v, P) + \frac{n}{2(k+1)} w(P_i)$ for any $0 \leq i \leq k$.

PROOF For any $v \in V_j \subseteq U_i$,

$$\begin{aligned} d_G(v, M) &\leq \frac{1}{2}(d_G(v, m_i) + d_G(v, m_{i+1})) \\ &\leq \frac{1}{2}((d_Y(v, P) + d_Y(m_i, p_j)) + (d_Y(v, P) + d_Y(p_j, m_{i+1}))) \\ &= d_Y(v, P) + \frac{1}{2} d_Y(m_i, m_{i+1}) \\ &= d_Y(v, P) + \frac{1}{2} w(P_i). \end{aligned}$$

Since $|U_i| \leq \frac{n}{k+1}$, we have

$$\begin{aligned} \sum_{v \in U_i} d_G(v, M) &\leq \sum_{v \in U_i} (d_Y(v, P) + \frac{1}{2} w(P_i)) \\ &\leq \sum_{v \in U_i} d_Y(v, P) + \frac{n}{2(k+1)} w(P_i). \end{aligned}$$

□

The vertex set M is defined to partition the vertices into small pieces. Our goal is to correctly guess $m_1, m_2, \ldots, m_k$ and construct a tree X spanning these vertices along with s_1 and s_2, with the property that $d_X(v, s_1) + d_X(v, s_2) \leq w(P)$ for any $v \in V(X)$. Once such a tree X has been constructed, we extend it to a spanning tree T by adding the shortest-paths forest with vertices in $V(X)$ as the multiple roots. We shall first show that T is a $(\frac{k+2}{k+1})$-approximation, and the algorithm for constructing the tree X will be discussed later.

LEMMA 5.22
Let X be a tree spanning M and $d_X(v, s_1) + d_X(v, s_2) \leq w(P)$ for any $v \in V(X)$. The spanning tree T, which is the union of X and the shortest-paths forest with vertices in $V(X)$ as the multiple roots, is a $(\frac{k+2}{k+1})$-approximation of the 2-MRCT.

PROOF By definition, the vertices $m_1, m_2, \ldots, m_k$ partition the vertex set V into $(U, U_0, U_1, \ldots, U_k)$ and the subsets satisfy the properties in Lemmas 5.20 and 5.21. Since $d_X(v, s_1) + d_X(v, s_2) \leq w(P)$ for any $v \in V(X)$, we have

$$C_m(T, s_1, s_2)$$
$$= \sum_{v \in V} (d_T(v, s_1) + d_T(v, s_2))$$
$$\leq \sum_{v \in V} (2d_T(v, X) + w(P))$$
$$= n \times w(P) + 2 \sum_{v \in V} d_T(v, X)$$
$$\leq n \times w(P) + 2 \sum_{v \in U} d_T(v, X) + 2 \sum_{0 \leq i \leq k} \sum_{v \in U_i} d_T(v, X). \quad (5.19)$$

Since $d_T(v, X) = d_G(v, X) \leq d_G(v, M)$ for any $v \in U$, by Lemma 5.20, we have

$$\sum_{v \in U} d_T(v, X) \leq \sum_{v \in U} d_G(v, M) \leq \sum_{v \in U} d_Y(v, P). \quad (5.20)$$

Similarly, $d_T(v, X) \leq d_G(v, M)$ for any $v \in U_i$, and, by Lemma 5.21, we have

$$\sum_{0 \leq i \leq k} \sum_{v \in U_i} d_T(v, X) \leq \sum_{0 \leq i \leq k} \left(\sum_{v \in U_i} d_Y(v, P) + \frac{n}{2(k+1)} w(P_i) \right)$$
$$\leq \sum_{0 \leq i \leq k} \sum_{v \in U_i} d_Y(v, P) + \frac{n}{2(k+1)} w(P). \quad (5.21)$$

By (5.19), (5.20), and (5.21),

$$C_m(T, s_1, s_2) \leq n \times w(P) + 2 \sum_{v \in U} d_Y(v, P) + 2 \sum_{0 \leq i \leq k} \sum_{v \in U_i} d_Y(v, P) + \frac{n}{k+1} w(P)$$
$$\leq \frac{k+2}{k+1} n \times w(P) + 2 \sum_{v \in V} d_Y(v, P)$$
$$\leq \frac{k+2}{k+1} C_m(Y, s_1, s_2)$$

since $C_m(Y, s_1, s_2) = n \times w(P) + 2 \sum_{v \in V} d_Y(v, P)$ by Lemma 5.18. □

The PTAS is listed below.

Algorithm: 2MRCT_PTAS
Input: A graph $G = (V, E, w)$, $s_1, s_2 \in V$, and an integer $k \geq 0$.

Optimal Communication Spanning Trees

Output: A spanning tree T of G.
For each k-tuple $(m_1, m_2, \ldots, m_k)$ of not necessarily distinct vertices, use the following steps to find a spanning tree T, and output the tree of minimal routing cost.

1: Let $m_0 = s_1$ and $m_{k+1} = s_2$.
2: Find a tree X by the following substeps:
2.1: Initially X contains only one vertex m_0. /* $\delta = 0$. */
2.2: for $i = 0$ to k do
2.3:   Find any shortest path Q from m_i to m_{i+1}.
       Let $Q = (q_0 = m_i, q_1, \ldots, q_h = m_{i+1})$.
2.4:   for $j = 0$ to $h - 1$ do
       Let $\bar{X} = X$.
2.5:   Add edge (q_j, q_{j+1}) to X. /* $\delta = \delta + w(q_j, q_{j+1})$. */
2.6:   if X contains a cycle $(a_0 = q_{j+1}, a_1, a_2, \ldots, q_j, a_0)$ then
2.7:   [Case 1.] $q_{j+1} \in V(SP_{\bar{X}}(s_1, q_j))$:
             Delete the edge (a_b, a_{b+1}) such that both $d_X(a_0, a_b)$ and $d_X(a_{b+1}, a_0)$ are no more than one half of the cycle length.
2.8:   [Case 2.] $q_{j+1} \notin V(SP_{\bar{X}}(s_1, q_j))$: Delete edge (a_0, a_1).
3: Find the shortest-paths forest spanning $V(G)$ with all vertices in $V(X)$ as the multiple roots.
4: Let T be the union of the forest and X.

The properties of X constructed at Step 2 are shown in the next lemma.

LEMMA 5.23

Let $m_1, m_2, \ldots, m_k$ be k vertices such that P connects the consecutive m_i. The graph X constructed at Step 2 is a tree and $d_X(v, s_1) + d_X(v, s_2) \leq w(P)$ for any $v \in V(X)$. Furthermore, it takes $O(kn^2)$ to construct X.

PROOF Starting from a single vertex, X is repeatedly augmented edge by edge. As in the comment of Step 2.5, let δ be the total weight of all edges that have been added to X so far. When X is completed,

$$\delta = \sum_i SP_G(m_i, m_{i+1}) \leq \sum_i w(P_i) = w(P).$$

It is sufficient to show that the following two properties are kept at the end of Step 2.8.

- X is a tree.

- $d_X(v, s_1) + d_X(v, q_{j+1}) \leq \delta$ for any vertex $v \in V(X)$, where (q_j, q_{j+1}) is the last edge added to X.

Initially the properties are true since X contains only one vertex. To avoid confusion, let $\bar{X}$ denote the constructed tree before edge (q_j, q_{j+1}) is added,

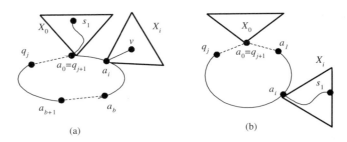

FIGURE 5.9: Removing an edge from a cycle.

and X denote the tree at the end of Step 2.8. Similarly let $\bar{\delta}$ and δ be the value before and after adding the edge respectively. It is obvious that X remains a tree if $\bar{X}$ is a tree, since it is connected and contains no cycle. If the edge does not cause a cycle, the second property is also straightforward. Now we consider that there is a cycle $(a_0 = q_{j+1}, a_1, a_2, \ldots, q_j, a_0)$. Let $X_i \subset V(X)$ denote the set of vertices which are connected to the cycle at a_i (Figure 5.9).

Suppose that, for any vertex $v \in V(\bar{X})$,

$$d_{\bar{X}}(v, s_1) + d_{\bar{X}}(v, q_j) \leq \bar{\delta}. \tag{5.22}$$

We shall show that

$$d_X(v, s_1) + d_X(v, q_{j+1}) \leq \bar{\delta} + w(q_j, q_{j+1}) = \delta. \tag{5.23}$$

There are two cases.

Case 1: $q_{j+1} \in V(SP_{\bar{X}}(s_1, q_j))$, i.e., $s_1 \in X_0$ (Figure 5.9(a)). First we show that the edge (a_b, a_{b+1}) always exists and can be found by traveling the cycle. Starting from a_0, we travel the cycle and compute the distance from a_0. We can always find vertex a_b such that $w(a_0, a_1, \ldots, a_b) \leq \frac{L}{2}$ and $w(a_0, a_1, \ldots, a_b, a_{b+1}) > \frac{L}{2}$, where L is the cycle length. After removing edge (a_b, a_{b+1}), $d_X(a_0, a_b) \leq \frac{L}{2}$ and $d_X(a_{b+1}, a_0) < \frac{L}{2}$. It implies that, for any vertex a_i on the cycle,

$$d_X(a_i, q_{j+1}) \leq \frac{L}{2}. \tag{5.24}$$

If $v \in X_0$, both the distances from v to s_1 and q_{j+1} do not change, and $d_X(v, q_{j+1}) \leq d_{\bar{X}}(v, q_j)$. Therefore (5.23) is true. Otherwise, assume $v \in X_i$. $d_{\bar{X}}(v, s_1) = d_{\bar{X}}(v, a_i) + d_{\bar{X}}(a_i, q_{j+1}) + d_{\bar{X}}(q_{j+1}, s_1)$, and $d_{\bar{X}}(v,q_j) = d_{\bar{X}}(v,a_i) + d_{\bar{X}}(a_i,q_j)$. By (5.22), we have

$$d_{\bar{X}}(v,s_1) + d_{\bar{X}}(v,q_j) \leq \bar{\delta},$$
$$2d_{\bar{X}}(v,a_i) + d_{\bar{X}}(a_i,q_{j+1}) + d_{\bar{X}}(q_{j+1},s_1) + d_{\bar{X}}(a_i,q_j) \leq \bar{\delta},$$
$$2d_{\bar{X}}(v,a_i) + (L - w(q_j,q_{j+1})) + d_{\bar{X}}(q_{j+1},s_1) \leq \bar{\delta},$$
$$2d_{\bar{X}}(v,a_i) + L + d_{\bar{X}}(q_{j+1},s_1) \leq \delta.$$

By (5.24),

$$d_X(v,s_1) + d_X(v,q_{j+1}) = 2d_X(v,a_i) + 2d_X(a_i,q_{j+1}) + d_X(q_{j+1},s_1)$$
$$\leq 2d_{\bar{X}}(v,a_i) + L + d_{\bar{X}}(q_{j+1},s_1) \leq \delta.$$

Case 2: $q_{j+1} \notin V(SP_{\bar{X}}(s_1,q_j))$ (Figure 5.9(b)). In this case edge (a_0,a_1) is removed. If $v \notin X_0$, $d_X(v,s_1) = d_{\bar{X}}(v,s_1)$ and $d_X(v,q_{j+1}) = d_{\bar{X}}(v,q_j) + w(q_j,q_{j+1})$. Then (5.23) follows (5.22). Otherwise $v \in X_0$. Assume $s_1 \in X_i$. $d_{\bar{X}}(v,s_1) = d_{\bar{X}}(v,a_0) + d_{\bar{X}}(a_0,a_i) + d_{\bar{X}}(a_i,s_1)$, and $d_{\bar{X}}(v,q_j) = d_{\bar{X}}(v,a_0) + d_{\bar{X}}(a_0,q_j)$. By (5.22),

$$d_{\bar{X}}(v,s_1) + d_{\bar{X}}(v,q_j) \leq \bar{\delta},$$
$$2d_{\bar{X}}(v,a_0) + d_{\bar{X}}(a_i,s_1) + d_{\bar{X}}(a_0,a_i) + d_{\bar{X}}(a_0,q_j) \leq \bar{\delta},$$
$$2d_{\bar{X}}(v,a_0) + d_{\bar{X}}(a_i,s_1) + d_{\bar{X}}(a_0,a_i) + L \leq \delta.$$

The last step is obtained by $d_{\bar{X}}(a_0,q_j) + w(q_j,q_{j+1}) = L$. We have

$$d_X(v,s_1) + d_X(v,q_{j+1}) = 2d_X(v,a_0) + d_X(a_0,a_i) + d_X(a_i,s_1)$$
$$\leq 2d_{\bar{X}}(v,a_0) + L + d_{\bar{X}}(a_i,s_1) \leq \delta.$$

We have shown (5.23) for both cases. By induction, $d_X(v,s_1) + d_X(v,s_2) \leq w(P)$ for any $v \in V(X)$ at the end of Step 2. The number of vertices on $SP_G(m_i, m_{i+1})$ is at most $O(n)$. For each vertex on the path, we check if it causes a cycle and remove an edge from the cycle if it exists. All these can be done in $O(n)$ time. Consequently the time complexity is $O(kn^2)$. □

The main result of this section is concluded in the next theorem.

THEOREM 5.7
The 2-MRCT problem admits a PTAS. For any constant $\varepsilon > 0$, a $(1+\varepsilon)$-approximation of a 2-MRCT of a graph G can be found in polynomial time. The time complexity is $O(n^{\lceil 1/\varepsilon+1\rceil})$ for $0 < \varepsilon < 1$, and $f_{SP}(G)$ for $\varepsilon = 1$.

PROOF For any $\varepsilon > 0$, we choose an integer $k = \lceil \frac{1}{\varepsilon} - 1 \rceil$ and run Algorithm 2MRCT_PTAS. For $\varepsilon \geq 1$, i.e., $k = 0$, the algorithm is equivalent to Algorithm 2MRCT_1 and finds a 2-approximation in $f_{SP}(G)$ time.

For $0 < \varepsilon < 1$, i.e., $k \geq 1$, the algorithm tries each possible k-tuple in each iteration. There exists a k-tuple $(m_1, m_2, \ldots, m_k)$ which partitions the vertex set V into $(U, U_0, U_1, \ldots, U_k)$ and the subsets satisfy the properties in Lemmas 5.20 and 5.21. By Lemma 5.23 and Steps 3 and 4, the output T is a tree. Also by Lemma 5.23, $d_X(v, s_1) + d_X(v, s_2) \leq w(P)$ for any $v \in V(X)$. Then, by Lemma 5.22, the spanning tree T constructed at Step 4 is a $(\frac{k+2}{k+1})$-approximation of the 2-MRCT. The number of all possible k-tuples is $O(n^k)$ for constant k. Similar to the proof of Lemma 5.19 and by Lemma 5.23, each iteration takes $O(kn^2)$ time. The time complexity of the algorithm is therefore $O(n^{k+2})$. That is, the ratio is $\frac{k+2}{k+1} = 1 + \varepsilon$, and the time complexity is $O(n^{\lceil 1/\varepsilon + 1 \rceil})$, which is polynomial for constant ε. □

5.5 Applications

The problems discussed in this section are generalizations of the MRCT problem. Similar to the MRCT problem, one of the applications of the problems is in network design. Suppose that each vertex v represents a city and $r(v)$ is the population of the city. It is reasonable to assume that the communication requirement of a pair of vertices is proportional to the product of the populations of the two cities. The PROCT is the tree structure with minimum communication cost.

The SROCT problem may arise in the following situation: For each node in the network, there is an individual message to be sent to every other node and the amount of the message is proportional to the weight of the receiver. With this assumption, the communication cost of a spanning tree T is $\sum_{u,v} r(v) d_T(u, v)$, which is exactly one half of $C_s(T)$.

In a computer network, most of the communications are between servers and clients. In general, there are only few servers and many clients. The study of the p-source MRCT problem attempts to provide solutions for such applications.

Similar to that the MRCT finds applications in the SP-alignment, the generalizations of the MRCT may be helpful for the generalized SP-alignment. A simple generalization of the SP objective for multiple alignments is to weight the different sequence pairs in the alignment differently in the objective function. Given a priority value r_{ij} for the pair i, j of sequences, the *generalized sum-of-pairs* objective for multiple alignment is to minimize the sum, over all pairs of sequences, of the pairwise distance between them in the alignment multiplied by the priority value of the pair. This allows one to increase the priority of aligning some pairs while down-weighing others, using other information (such as evolutionary relationship) to decide on the priorities. An extreme case of assigning priorities is the *threshold* objective.

In an evolutionary context, a multiple alignment is used to reconstruct the blocks or motifs in a single ancestral sequence from which the given sequences have evolved. However, if the evolutionary events of the ancestral sequence occur randomly at a certain rate over the course of time, and independently at each location (character) of the string, after a sufficiently long time, the mutated sequence appears essentially like a random sequence compared to the initial ancestral sequence. If we postulate a threshold time beyond which this happens, this translates roughly to a threshold edit distance between the pair of sequences. The threshold objective sets r_{ij} to be one for all pairs of input sequences whose edit-distance is less than this threshold and zero for other pairs which are more distant. In this way we try to capture the most information about closely related pairs in the objective function by setting an appropriate threshold.

In the same vein as Gusfield [48], a δ-approximation of the OCT can be used to approximate the generalized SP objective. Let d_{ij} denote the edit distance between sequences i and j. The theorem guarantees a tree whose communication cost using the r_{ij} values given by the priority function is at most δ times $\sum_{i,j} r_{ij} d_{ij}$, which is a lower bound on the generalized SP value of any alignment. The Feng-Doolittle procedure guarantees that the generalized SP value of the resulting alignment is at most the communication cost of the tree which in turn is at most δ times the generalized SP value of any alignment.

5.6 Summary

In this chapter, we introduce the optimal communication spanning tree problem. Several restricted versions of the problem are investigated. Some of the problems are also generalizations of the minimum routing cost spanning tree problem. For each problem, we discuss its computational complexity and approximation algorithms. Table 5.1 summarizes the results. For the approximation ratio of OCT problem, see the Bibliographic Notes.

Bibliographic Notes and Further Reading

The current best approximation ratio for the general OCT problem is due to Yair Bartal's algorithms which approximate arbitrary metrics by tree metrics. He first presented a randomized algorithm [8] and then derandomized it to a deterministic algorithm [9]. Its application to approximating the OCT

TABLE 5.1: The objectives and currently best ratios of the OCT problems.

Problem	Objective	Ratio
OCT	$\sum_{u,v} \lambda(u,v) d_T(u,v)$	$O(\log n \log \log n)$
PROCT	$\sum_{u,v} r(u) r(v) d_T(u,v)$	PTAS
SROCT	$\sum_{u,v} (r(u) + r(v)) d_T(u,v)$	2
MRCT	$\sum_{u,v} d_T(u,v)$	PTAS
p-MRCT	$\sum_{u \in S} \sum_{v \in V} d_T(u,v)$	2
2-MRCT	$\sum_v (d_T(s_1, v) + d_T(s_2, v))$	PTAS

problem was pointed out in [100].

The PROCT and SROCT problems were introduced in [96]. In that paper, Bang Ye Wu et al. gave a 1.577-approximation algorithm for the PROCT problem and a 2-approximation algorithm for the SROCT problem. The PTAS using the Scaling-and-Rounding technique for a PROCT problem was presented in [98] by the same authors. Scaling the input instances is a technique that has been used to balance the running time and the approximation ratio. For example, Oscar H. Ibarra and Chul E. Kim used the scaling technique to develop a *fully polynomial time approximation scheme* (FPTAS) for the knapsack problem [58], and some improvement was made by Eugene L. Lawler [70]. A nice explanation of the technique can also be found in [43](pp. 134–137).

The time complexity in Corollary 5.4 is not currently the best result. There is a more efficient 1.577-approximation algorithm for the MRCT. In [96], it was shown that the time complexity can be reduced to $O(n^3)$ with the same approximation ratio. See Exercises 5-6 through 5-9.

The NP-hardness of the 2-MRCT was shown by Bang Ye Wu [93], in which the reduction is from the EXACT COVER BY 3-SETS (X3C) problem ([SP2] in [43]). The reduction in Section 5.4.1 is from [94]. The transformation is simpler and easier to extend to the weighted case, which is designed to show the NP-hardness of the p-MRCT problem for any fixed p. A similar reduction (for 2-MRCT) was also shown by Harold Connamacher and Andrzej Proskurowski [23]. They showed that the 2-MRCT problem is NP-hard in the strong sense. The PTAS for the 2-MRCT problem also appeared in [93]. In addition to the PTAS for the 2-MRCT problem, there is also a PTAS for the weighted 2-MRCT problem. But the PTAS works only for metric inputs and the counterpart on general graphs was left as an open problem.

Exercises

5-1. Let $T = (V, E, w)$ be a tree with $V = \{v_i | 1 \leq i \leq n\}$ and $r(v_i) = i$.

  (a) What is the p.r.c. cost if $T = (v_1, v_2, \ldots, v_n)$ is a path and each edge has unit length?

  (b) What is the p.r.c. cost if T is a star centered at v_1 and $w(v_1, v_i) = i$ for each $1 < i \leq n$?

5-2. Find the s.r.c. cost for each of the two trees in the previous problem.

5-3. What is the r-centroid of the path in Exercise 5-1?

5-4. Show Lemma 5.1.

5-5. Prove Fact 5.2 in Section 5.2.4.

5-6. Show the following property.

  Let $T = \text{Tstar}(x, y, X, Y)$ be a minimum routing cost 2-star of a metric graph $G = (V, E, w)$. For any $u \in X$ and $v \in Y$, $w(x, u) - w(y, u) \leq w(x, v) - w(y, v)$.

5-7. Given two specified vertices x and y of a metric graph, design an algorithm for finding the vertex partition (X, Y) minimizing the routing cost of $\text{Tstar}(x, y, X, Y)$. (Hint: Define a function $f(v) = w(x, v) - w(y, v)$ for all $v \in V$ and use the property in Exercise 5-6.)

5-8. Give an $O(n^3 \log n)$-time 1.577-approximation algorithm for the MRCT problem by using the result in Exercise 5-7.

5-9. The following property was shown in [96].

  For a metric graph G, there exists a 1.577-approximation solution Y of the MRCT problem on G such that Y is either a 1-star or a 2-star with $0.366n$ leaves connected to one of its internal nodes.

  Design an $O(n^3)$-time 1.577-approximation algorithm for the MRCT by using the property and a technique similar to the one used in Exercise 5-7.

5-10. Let $G = (V, E, w)$ and $U \subset V$. The Steiner MRCT problem asks for a tree T minimizing the total distance among all vertices in U, i.e., we want to minimize $\sum_{u,v \in U} d_T(u, v)$. Show that the Steiner MRCT problem is a special case of the PROCT problem.

5-11. Show that the Steiner MRCT problem on general graphs can be reduced to its metric version. (Hint: Use the result in Section 4.4.)

5-12. Design a PTAS for the Steiner MRCT problem by slightly modifying the PTAS for the MRCT in Chapter 4. What is the time complexity of the PTAS?

5-13. The objective function for the SROCT problem is given by $\sum_{u,v}(r(u)+r(v))d_T(u,v)$. Show that the problem is equivalent if the objective function is given by $\sum_{u,v}r(u)d_T(u,v)$.

5-14. Find the routing cost for the following trees with given sources.

(a) $T=(v_1,v_2,\ldots,v_n)$ is a path and each edge has unit length. The sources are given by $S=\{v_{ik}|1\leq i\leq p\}$ and $n=kp$, in which k is a positive integer.

(b) T is a complete binary tree with $|V(T)|=2^k-1$ and each edge has unit length. All the leaves are source vertices.

5-15. Let s be the given source in a 1-MRCT problem. Show that any shortest-paths tree rooted at s is an optimal solution.

5-16. Show Corollary 5.6.

Chapter 6

Balancing the Tree Costs

6.1 Introduction

With different aspects, there are different measurements of the goodness of a spanning tree. For a nonnegative edge-weighted graph G, the weight on each edge represents the distance and reflects both the cost to install the link (building cost) and the cost to traverse it after the link is installed (routing cost). When considering the building cost only, we are looking for the spanning tree with minimum total weight of edges. A minimum spanning tree (MST) is an optimal solution under the consideration. When considering the routing cost only, a p-source minimum routing cost spanning tree (p-MRCT) is an optimal solution for the case of p sources. Note that a 1-MRCT is just a shortest-paths tree and a minimum routing cost spanning tree (MRCT) is for the case that all vertices are sources.

Usually, removing any edge from a graph will increase the routing cost but decrease the building cost. Therefore graphs with smaller routing costs might require larger building costs, and we often need to make a tradeoff between the two costs. In this chapter, we shall discuss problems concerning both costs. The problems of constructing the following two spanning trees are discussed in this chapter, in which the input $G = (V, E, w)$ is a simple connected undirected graph.

1. LIGHT APPROXIMATE SHORTEST-PATHS TREES (LAST): A LAST is a spanning tree with total edge weight no more than a constant times that of an MST and the distance from every node to the specified source no more than a constant times the shortest path length. Formally, let $\alpha, \beta \geq 1$ be two real numbers. For a graph $G = (V, E, w)$ and $s \in V$, a spanning tree T is an (α, β)-LAST of G if $d_T(s, v) \leq \alpha d(s, v)$ for each $v \in V$ and $w(T) \leq \beta w(MST(G))$.

2. LIGHT APPROXIMATE ROUTING COST SPANNING TREES (LART): A LART is a spanning tree with total edge weight no more than a constant times that of a MST and the routing cost no more than a constant times the cost of an MRCT. Formally, let $\alpha, \beta \geq 1$ be two real numbers. For a graph $G = (V, E, w)$, a spanning tree T is an (α, β)-LART of G if $C(T) \leq \alpha C(MRCT(G))$ and $w(T) \leq \beta w(MST(G))$.

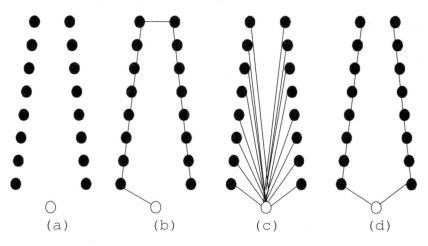

FIGURE 6.1: A minimum spanning tree, a shortest-paths tree, and an LAST of a set of vertices in Euclidean space.

6.2 Light Approximate Shortest-Paths Trees
6.2.1 Overview

A minimum spanning tree is a connected subgraph with minimum building cost (total edge weight) but may suffer large routing cost. Meanwhile, in the case of single source, a shortest-paths tree is a subgraph with minimum routing cost but may require large building cost. A LAST is a tree balancing the two costs. The following example illustrates the three spanning trees of vertices in Euclidean space.

Example 6.1
Figure 6.1(a) shows a set of vertices in the Euclidean space; (b) and (c) show an MST and a shortest-paths tree respectively; and (d) shows an LAST. On the MST, there are some vertices with long paths to the source, while the shortest-paths tree has a tremendous total edge weight. Consider the LAST shown in (d). It improves the long path problem in the MST by increasing the total edge weight slightly. □

In the case that the underlying graph is an unweighted graph, any shortest-paths tree rooted at the source is a (1,1)-LAST since the total edge weight is $(n-1)$, the same as any spanning tree, and the distance from the source to any vertex is the shortest path length. The problem is more interesting and important when the underlying graph has weights on the edges. In this section, we shall present an algorithm for constructing an LAST of a weighted graph and its performance analysis.

6.2.2 The algorithm

The input is an undirected graph $G = (V, E, w)$, a source $s \in V$ and an $\alpha > 1$, in which G is connected and $w : E \to Z_0^+$ is the edge weight function. The algorithm constructs an $(\alpha, 1 + 2(\alpha - 1))$-LAST rooted at s. First it constructs a minimum spanning tree T_M and a shortest-paths tree T_S rooted at s. The basic idea of the algorithm is to traverse the minimum spanning tree by performing a depth-first search. While traversing, it maintains a current tree and the estimate of the distance from the root to each vertex. As a vertex is encountered, it checks the distance estimate to the root to ensure the distance requirement for that vertex in the current tree. Once the distance requirement is violated, the edges of the shortest path between the vertex and the root are added into the current tree and other edges are discarded so that a tree structure is maintained. After all vertices have been visited, the remaining tree is the desired LAST. The final tree is not too heavy because a shortest path is only added if the replaced path is heavier by a factor of α. This allows a charging argument bounding the net weight of the added paths.

A depth-first search for a tree is a tour traversing each edge exactly twice: once in each direction. It is reasonable to represent the walk as a *Eulerian cycle*. In general, for an undirected graph (not necessary simple), a Eulerian cycle is a cycle traversing each edge exactly once. A necessary and sufficient condition for the existence of a Eulerian cycle in a graph is that each vertex has even degree. A graph is a *Eulerian graph* if there exists a Eulerian cycle. For any tree, by doubling each edge we can convert the tree into a Eulerian graph (a doubling tree), and a Eulerian cycle on the graph corresponds to a depth-first search on the tree.

Example 6.2
Figure 6.2 illustrates an example: (a) is a tree rooted at S; and (b) shows a Eulerian cycle. The cycle is $(S, A, B, A, C, A, S, D, E, D, S, F, G, F, H, F, S)$.
◻

Algorithm: FIND-LAST
Input: A graph $G = (V, E, w)$, a source $s \in V$ and $\alpha > 1$.
Output: An $(\alpha, 1 + 2(\alpha - 1))$-LAST rooted at s.
1:    Find a minimum spanning tree T_M and root T_M at s.
2:    Find a shortest-paths tree T_S rooted at s.
3:    Set $\delta[v] \leftarrow \infty$ and $\pi[v] \leftarrow$ NIL for each $v \in V$, and $\delta[s] \leftarrow 0$.
4:    Find a Euler cycle $(s = v_0, v_1, \ldots, v_{2n-2} = s)$ on the doubling tree of T_M.

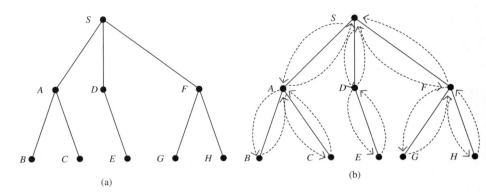

FIGURE 6.2: The Eulerian cycle on a doubling tree.

5: for $i \leftarrow 1$ to $2n-2$ do
   RELAX(v_{i-1}, v_i);
   if $\delta[v_i] > \alpha d(s, v_i)$ then
       ADD-PATH(v_i).
6: Output tree T with edge set $\{(v, \pi[v]) | v \in V - \{s\}\}$.

SUBROUTINE RELAX(u, v)
   if $\delta[v] > \delta[u] + w(u, v)$ then
       $\delta[v] \leftarrow \delta[u] + w(u, v)$;
       $\pi[v] \leftarrow u$;

SUBROUTINE ADD-PATH(u)
/*add the shortest path from v_i to s into the current tree */
   Let parent(u) be the parent of u in T_S.
   if $\delta[u] > d_G(s, u)$ then
       $\pi[u] \leftarrow$ parent(u);
       $\delta[u] \leftarrow d_G(s, u)$;
       ADD-PATH(parent(u));

The current tree is maintained by keeping a parent pointer $\pi[v]$ for each nonroot vertex v. For each vertex, a distance estimate to the root is stored at $\delta[v]$. This distance estimate, which is an upper bound on the true distance in the current tree, is used in deciding whether to add a path to the vertex. The algorithm builds and updates the tree and maintains the distance estimates by a sequence of calls to RELAX(u, v). The subroutine RELAX(u, v) checks if there is a shorter path from s to v via u. If so, the algorithm updates the distance estimate and the parent pointer.

When vertex v is visited, if $\delta[v]$ exceeds α times the distance from the

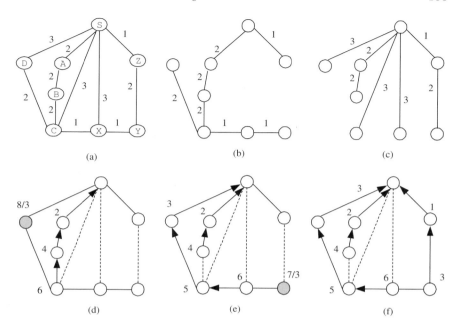

FIGURE 6.3: A sample execution of algorithm FIND-LAST.

root to v in the shortest-paths tree, then by calling ADD-PATH the edges of the shortest path are added to lower $\delta[v]$ to the shortest-path distance. The subroutine ADD-PATH is a recursive procedure which traverses the shortest path from v toward s, and adds the path edges by changing the parent pointer until it reaches a vertex from which a shortest path to s is already in the current tree ($\delta[u] = d_G(u, s)$).

Example 6.3
Figure 6.3 shows a sample execution of the algorithm with $\alpha = 2$. (a) is the given graph. (b) and (c) are a minimum spanning tree T_M and a shortest-path tree T_S respectively. Initially all parent pointers are NIL and each $\delta[v]$ is infinite. A Eulerian cycle on the T_M is found to be

$$(S, A, B, C, D, C, X, Y, X, C, B, A, S, Z, S).$$

The algorithm executes as follows:

1. Traversing along the Eulerian cycle from S to A, B, and arriving at C: For these nodes, the distances to the source do not exceed the requirement. Therefore the edges of T_M are used, and $\delta[A] = 2$, $\delta[B] = 4$ and $\delta[C] = 6$.

2. Traversing edge (C, D): the edge is first added ($\pi[D] = C$) but the algorithm finds that

$$\delta[D] = 8 > \alpha \times d(D, S) = 2 \times 3 = 6,$$

as in Frame (d). The shortest path from D to S is then added by changing $\pi[D]$ to S. Also $\delta[D] \leftarrow 3$. Note that since $\pi[D]$ is changed to S, the previous edge (C, D) is removed and a tree structure is maintained. Frame (d) shows the status of this step.

3. Traversing from D back to C: RELAX(D, C) is called and the algorithm finds that vertex D provides a shorter path (with length 5) to S. The parent pointer $\pi[C]$ is replaced with D and $\delta[C] \leftarrow 5$.

4. Traversing from C to X and then to Y: The edge (C, X) is added ($\pi[X] = C$, $\delta[X] = 6$). When Y is visited, the distance requirement is violated since $7 > 2 \times 3$, as in Frame (e), and the shortest path to Y is added by setting $\pi[Y] \leftarrow Z$ and $\pi[Z] \leftarrow S$. Note that X's parent is also changed when RELAX(Y, X) is called. This was the final change made to the tree. Subsequent traversal had no effect, and the result is given in Frame (f).

◻

6.2.3 The analysis of the algorithm

Let T be the tree constructed by algorithm FIND-LAST. We are going to show that T is a $(\alpha, 1 + 2/(\alpha - 1))$-LAST. First, since a shortest path is added whenever $\delta[v] > \alpha d(s, v)$, the next result is trivial.

FACT 6.1
For each $v \in V$, $d_T(s, v) \leq \alpha d(s, v)$.

An important feature of the algorithm is that the total weight of T is not too large.

LEMMA 6.1
The total weight of T is at most $(1 + 2/(\alpha - 1))w(T_M)$.

PROOF We show the lemma by giving an amortized analysis. Define the potential function Φ to be the distance estimate of the current vertex. Φ increases when an edge of T_M is added, and decreases when the shortest path is added. When an edge (u, v) of T_M is added, the increment is the edge weight $w(u, v)$. Hence the total increment is at most $2w(T_M)$ since each edge of T_M is traversed twice.

When a shortest path from s to v is added, Φ is lowered to $d(s,v)$. Since the shortest path is added only when $\delta[v] > \alpha d(s,v)$, Φ decreases at least $(\alpha-1)d(s,v)$. Therefore the total weight of the added shortest path is bounded by the total decrement to Φ divided by $(\alpha - 1)$. Since Φ is initially 0 and always nonnegative, the total decrement is bounded by the total increment. Hence we have that the total weight of all added shortest paths is at most $2/(\alpha - 1)$ times $w(T_M)$, and

$$w(T) \leq w(T_M) + \frac{2}{(\alpha - 1)} w(T_M) = \left(1 + \frac{2}{(\alpha - 1)}\right) w(T_M).$$

☐

Next we consider the time complexity of the algorithm.

LEMMA 6.2

If a minimum spanning tree and a shortest-paths tree are given, the algorithm FIND-LAST *runs in $O(n)$ time, where n is the number of vertices.*

PROOF First a Eulerian cycle can be found in linear time by performing a depth-first search in the tree. Since each edge of T_M is traversed twice and the number of edges is $(n-1)$, the total time complexity of executing RELAX is $O(n)$. Although there may be several edges involved in ADD-PATH and the subroutine may be called several times, each edge of the shortest-paths tree is added at most once. Therefore the total time complexity of executing ADD-PATH is also $O(n)$. In summary, the time complexity of the algorithm is $O(n)$ if a minimum spanning tree and a shortest-paths tree are given. ☐

In the case that T_M and T_S are not given, they can be constructed in $O(m + n \log n)$ time, where m is the number of edges of the graph. We summarize the result of this section in the following theorem.

THEOREM 6.1

The algorithm FIND-LAST *constructs a $(\alpha, 1 + 2/(\alpha - 1))$-LAST of a graph in $O(m + n \log n)$ time. If a minimum spanning tree and a shortest-paths tree are given, the time complexity is $O(n)$.*

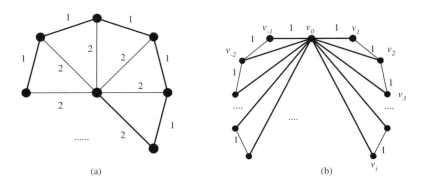

FIGURE 6.4: (a) A simple example of an MST with large routing cost: The routing cost of the MST is $\Theta(n)$ times that of the star. (b) A small routing cost tree with large weight. The vertices are labelled by $v_{-n/2},\ldots,v_{-1}, v_0, v_1, v_2,\ldots,v_{n/2-1}$. $w(v_i, v_{i+1}) = 1$ and $w(v_0, v_i) = (|i| + 2)/3$ for each i. The routing cost of the star is about one half that of an MST (a path in this example), while the weight of the star is $\Theta(n)$ times that of an MST.

6.3 Light Approximate Routing Cost Spanning Trees

6.3.1 Overview

While an LAST balances the building cost and the routing cost from one source vertex to the others, an LART balances the building cost and the all-to-all routing cost. As we have seen in Example 4.2, the routing cost of a tree is determined not only by the weight of edges but also by the topology. Although both the minimum spanning tree (MST) and the minimum routing cost spanning tree (MRCT) tend to use light edges, a tree with small weight may have a large routing cost and vice versa. For instance, we can easily construct a graph such that the routing cost of its MST is $\Theta(n)$ times the routing cost of its MRCT (Figure 6.4). Similarly, a spanning tree with a constant times the minimum routing cost may have a tree weight as large as $\Theta(n)$ times the weight of an MST.

As shown in Theorem 4.2, there is a vertex such that the routing cost of any shortest-paths tree rooted at the vertex is no more than twice of that of the whole graph. If the input graph G is an unweighted graph (each edge has weight one), we can easily find a shortest-paths tree such that its weight is $n-1$ and its routing cost is no more than twice of $C(G)$. Therefore we have the result stated in Fact 6.2.

FACT 6.2
For an unweighted graph, there is a shortest-paths tree which is a (2,1)-LART.

Similar to the LAST problem in the previous section, we shall focus on the case of weighted graphs.

We briefly describe the main ideas of finding an (α, β)-LART of a graph as follows. Recall that a *metric graph* is a complete graph in which the edge weights satisfy the triangle inequality.

1. We shall first focus on metric graphs. Then, similar to the MRCT problem, we show that the algorithm can be also applied to a general graph with arbitrary nonnegative distances.

2. A k-star is a spanning tree with at most k internal nodes. For a metric graph G, there exists a k-star whose routing cost is at most $(k+3)/(k+1)$ times the minimum. However, the weight of a minimum routing cost k-star may be large. For example, the weight of a minimum routing cost 1-star may be $\Theta(n)$ times $w(MST(G))$.

3. Consider the algorithm FIND-LAST in the previous section for constructing an $(\alpha, 1+2/(\alpha-1))$-LAST rooted at a vertex. If we check the LAST rooted at each vertex and choose the one with minimum routing cost, we can show that it is a $(2\alpha, 1+2/(\alpha-1))$-LART. To find an LART achieving more general tradeoff, we use general k-stars. Let R be a vertex set containing the k internal nodes of the k-star with approximate routing cost of the $MRCT(G)$. We construct a *light approximated shortest-path forest* with multiple roots R by the algorithm FIND-LAST. Then we connect the forest into a tree T by adding the edges of a minimum-weight tree spanning R. Although the routing cost of T can not be arbitrarily close to the optimal as a minimum k-star does, we can show that it also ensures a good ratio.

We shall show and analyze the algorithm in the following subsections. The extension to general graph inputs and to other similar problems will be discussed at the end of this section.

6.3.2 The algorithm

In this subsection, we present the algorithm for finding an LART and analyze its time complexity. The following definition is a variant of the light approximate shortest-paths tree. Recall that the distance $d(v, R)$ between a vertex and a vertex set is the minimum of the distance between v and any vertex in R.

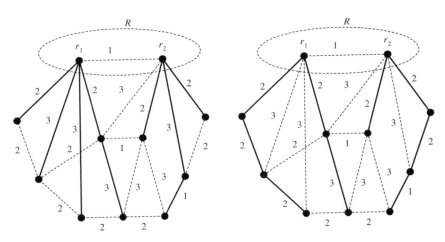

FIGURE 6.5: A shortest-paths forest and a light approximate shortest-paths forest.

DEFINITION 6.1 Let $G = (V, E, w)$ and $R \subset V$. For $\alpha \geq 1$ and $\beta \geq 1$, a light approximate shortest-paths forest (α,β)-LASF with roots R is a spanning forest F of G with $d_F(v, R) \leq \alpha \times d(v, R)$ for each $v \in V$ and $w(F) \leq \beta \times w(MST(G))$.

Example 6.4
An example of an LASF of a graph is given in Figure 6.5. Frame (a) is a shortest-paths forest with roots $R = \{r_1, r_2\}$ of a graph, in which $d_T(v, R) = d(v, R)$ for each vertex v. Frame (b) is a light approximate shortest-paths forest: $d_T(v, R) \leq 2d(v, R)$ for each vertex v. □

The algorithm FIND-LAST in the last section can be easily extended to the multiple roots variant. The method is just like the one for constructing a shortest-paths forest using the shortest-paths tree algorithm. First we create a dummy node and connect all roots to the dummy node by zero-length edges. Then we perform the algorithm FIND-LAST with the dummy node as the source. Removing the dummy node and the edges incident with the node, we obtain an LASF. It is easy to check that if the tree is an (α, β)-LAST, the forest is an (α, β)-LASF. For a metric graph, any edge is a shortest path between the two endpoints.

COROLLARY 6.1
Let G be a metric graph. For any k-vertex subset R of $V(G)$ and $\alpha > 1$, there

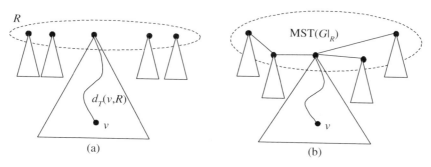

FIGURE 6.6: The tree constructed in one iteration of Algorithm FIND-LART. (a) A light approximate shortest-paths forest: For every vertex v, $d_T(v, R) \leq \alpha d_G(v, R)$. (b) Connecting the forest by the MST of R.

exists an algorithm which can construct an $(\alpha, 1 + 2/(\alpha - 1))$-LASF with roots R in $O(kn)$ time if $MST(G)$ is given.

The following is an algorithm for finding an LART, and an illustration is given in Figure 6.6.

Algorithm: FIND-LART
Input: A metric graph G, a real number $\alpha > 1$, and an integer k.
Output: An LART of G.
1: Find $MST(G)$.
2: For each $R \subset V(G)$ and $|R| \leq k$, use the following method to construct a spanning tree, and keep T with minimum $C(T)$.
2.1:   Find an $(\alpha, 1 + 2/(\alpha - 1))$-LASF T_1 with roots R.
2.2:   Find a minimum weight spanning tree T_0 of the induced subgraph $G|_R$.
2.3:   Set $T = T_0 \cup T_1$ and compute $C(T)$.
3: Output the tree T of minimum $C(T)$.

Let's first consider the time complexity of the algorithm. Since the input is a metric graph, there is no need to find the shortest-paths tree, and the minimum spanning tree can be found in $O(n^2)$ time. The time complexity is dominated by Step 2. For a specified k, the loop is executed $\binom{n}{1} + \binom{n}{2} + \ldots + \binom{n}{k} = O(n^k)$ times. For one iteration, an LASF T_1 takes $O(kn)$ time (Corollary 6.1) and T_0 takes $O(k^2)$ time since it spans only k vertices. Therefore the time complexity is $O(n^k(kn) + n^2)$, or $O(n^{k+1})$ for any constant $k \geq 1$.

LEMMA 6.3
The time complexity of FIND-LART is $O(n^{k+1})$.

6.3.3 The performance analysis

We now analyze the weight and the routing cost of the tree T constructed by algorithm FIND-LART. The bound of $w(T)$ is shown in the following lemma.

LEMMA 6.4
$w(T) \leq \left(f(k) + \frac{2}{\alpha-1}\right) w(MST(G))$, where $f(1) = 1$, $f(2) = 2$, and $f(k) = 3$ for $k > 2$.

PROOF By Theorem 6.1, when $k = 1$,

$$w(T) \leq \left(1 + \frac{2}{\alpha - 1}\right) w(MST(G)).$$

When $k = 2$, the tree T_0 contains only one edge. Since G is a metric graph, $w(T_0) \leq w(MST(G))$. We now consider the case where $k > 2$. A minimum weight tree spanning a subset of a graph is well known as a *Steiner minimal tree*. The weight of a Steiner minimal tree spanning R in G is no more than $w(MST(G))$. The ratio of the weight of the minimum tree without Steiner vertices to that of the Steiner minimal tree is known as the *Steiner ratio* in the literature. For a metric graph, the ratio is two [68]. Therefore $w(MST(G|_R)) \leq 2w(MST(G))$ for $k > 2$. □

In order to show the routing cost ratio, we recall some results introduced in Chapter 4.

- (Lemma 4.5) If S is a minimal δ-separator of $\widehat{T}$, then

$$C(\widehat{T}) \geq 2(1-\delta)n \sum_{v \in V} d_{\widehat{T}}(v, S) + 2\delta(1-\delta)n^2 w(S).$$

- (Lemma 4.11) For any constant $0 < \delta \leq 0.5$, and a spanning tree T of G, there exists a δ-spine Y of T such that $|CAL(Y)| \leq \lceil 2/\delta \rceil - 3$.

The following lemma will be used to show the routing cost ratio.

LEMMA 6.5
Let T be a spanning tree of a metric graph G. If S and Y are a minimal δ-separator and a δ-spine of T respectively, $\sum_{v \in V} d_G(v, CAL(Y)) \leq \sum_{v \in V} d_T(v, S) + (\delta n/4) w(S)$.

PROOF For the vertices v hung at a vertex in $CAL(Y)$,

$$d(v, CAL(Y)) \leq d_T(v, CAL(Y)) = d_T(v, S).$$

Let P be a δ-path in the spine with u_1 and u_2 as its endpoints. For any vertex v hung at P, by the triangle inequality,

$$d(v, CAL(Y)) \leq \min\{d_T(v, u_1), d_T(v, u_2)\} \leq d_T(v, S) + w(P)/2.$$

Since P is a δ-spine, there are at most $\delta n/2$ vertices hung at the internal vertices of P. Summing up the distances over all vertices, we have

$$\sum_{v \in V} d(v, CAL(Y)) \leq \sum_{v \in V} d_T(v, S) + \frac{\delta n}{4} \sum_{P \in Y} w(P)$$
$$= \sum_{v \in V} d_T(v, S) + \frac{\delta n}{4} w(S).$$

☐

Let T^* be the tree output by FIND-LART. We show the ratio of the routing cost in the next lemma.

LEMMA 6.6
Let $\widehat{T} = \mathrm{mrct}(G)$. $C(T^*) \leq \frac{k+3}{k+1} \alpha C(\widehat{T})$, for any integer $k \leq 6\alpha - 3$.

PROOF Let $\delta = 2/(k+3)$. By Lemma 4.11, there exists a δ-spine Y of $\widehat{T}$ such that $|CAL(Y)| \leq k$. Let T_1 be a light approximate shortest-paths forest with roots R and $T = T_1 \cup T_0$, in which $R = \{r_i | 1 \leq i \leq q\} = CAL(Y)$ and T_0 is a minimum weight spanning tree of the induced subgraph $G|_R$. Since $C(T^*) \leq C(T)$, we only need to prove the ratio of the routing cost of T. Let $V = V(G) = V(T)$. By Lemma 4.1 and that the routing load on any edge is no more than $n^2/2$,

$$C(T) \leq 2n \sum_{v \in V} d_T(v, R) + \sum_{e \in E(T_0)} l(T, e) w(e)$$
$$\leq 2n \sum_{v \in V} d_T(v, R) + \frac{n^2}{2} w(T_0). \tag{6.1}$$

Since T_1 is a light approximate shortest-paths forest,

$$\sum_{v \in V} d_T(v, R) \leq \alpha \sum_{v \in V} d(v, R). \tag{6.2}$$

By the inequalities (6.1) and (6.2), we have

$$C(T) \leq 2\alpha n \sum_{v \in V} d_G(v, R) + \frac{n^2}{2} w(T_0). \tag{6.3}$$

Let $S_1 = (R, E_S)$, in which the edge set E_S contains all the edges (u, v) if u and v are the two endpoints of some path in Y. By the triangle inequality, $w(S_1) \le w(S)$. Since S_1 is a spanning tree of the induced graph $G|_R$ and T_0 is a minimal one, $w(T_0) \le w(S_1) \le w(S)$. Then, by Lemma 6.5 and the inequality (6.3),

$$C(T) \le 2\alpha n \sum_{v \in V} d_{T^*}(v, S) + (\alpha\delta + 1)\frac{n^2}{2}w(S).$$

By Lemma 4.5,

$$C(T^*) \ge (1 - \delta)n \sum_{v \in V} d_{T^*}(v, S) + \delta(1 - \delta)n^2 w(S).$$

Comparing the two inequalities, we have

$$C(T) \le \max\{\frac{\alpha}{1-\delta}, \frac{\alpha\delta + 1}{4\delta(1-\delta)}\}C(T^*).$$

Note that $\alpha > 1$ and $0 < \delta \le 1/2$. Let

$$g(\delta) = \max\{\frac{\alpha}{1-\delta}, \frac{\alpha\delta + 1}{4\delta(1-\delta)}\}.$$

When $\delta \ge 1/(3\alpha)$,

$$\frac{\alpha}{1-\delta} \ge \frac{\alpha\delta + 1}{4\delta(1-\delta)},$$

and $g(\delta)$ decreases as δ decreases from $1/2$ to $1/(3\alpha)$. When $\delta \le 1/(3\alpha)$,

$$\frac{\alpha}{1-\delta} \le \frac{\alpha\delta + 1}{4\delta(1-\delta)},$$

and $g(\delta)$ increases as δ decreases from $1/(3\alpha)$. Therefore, $g(\delta)$ reaches its minimum when $\delta = 1/(3\alpha)$. Since $\delta = 2/(k+3)$, $C(T) \le \frac{k+3}{k+1}\alpha C(T^*)$, for any $k \le 6\alpha - 3$. □

In summary, the performance analysis of the algorithm is given in the following theorem.

THEOREM 6.2
Given a metric graph G, a $\left(\frac{k+3}{k+1}\alpha, \left(f(k) + \frac{2}{\alpha - 1}\right)\right)$-LART can be constructed in $O(n^{k+1})$ time for any real number $\alpha > 1$ and an integer $1 \le k \le 6\alpha - 3$, where $f(1) = 1$, $f(2) = 2$, and $f(k) = 3$ for $k > 2$.

6.3.4 On general graphs

We now extend the algorithm FIND-LART to the case where the input is a general graph.

Given a spanning tree T of the metric closure of a graph G, it is shown in Theorem 4.10 that, in $O(n^3)$ time, we can transform T into a spanning tree Y of G such that $C(Y) \leq C(T)$. By observing that $w(Y) \leq w(T)$ in the process of the construction, we may have the following corollary.

COROLLARY 6.2
Given an (α,β)-LART of the metric closure of a graph G, an (α,β)-LART of G can be constructed in $O(n^3)$ time.

Therefore, to find an LART of a general graph, we can first compute its metric closure, and then find an LART of the metric graph. Finally, we transform the tree into the desired LART of the original graph.

COROLLARY 6.3
Given a graph G, a $\left(\frac{k+3}{k+1}\alpha, \left(f(k) + \frac{2}{\alpha-1}\right)\right)$-LART can be constructed in $O(n^{k+1} + n^3)$ time for any real number $\alpha > 1$ and an integer $1 \leq k \leq 6\alpha - 3$, where $f(1) = 1$, $f(2) = 2$, and $f(k) = 3$ for $k > 2$.

6.4 Applications

There are numerous applications of an LAST and an LART in network design. While there is only one source in an LAST, all vertices are regarded as sources in an LART. The algorithm for constructing an LART can be applied to the following related problems. Since the MRCT problem is NP-hard, it is easy to see that these problems are also NP-hard. All the results can be obtained directly by Theorem 6.2.

1. The weight-constrained minimum routing cost spanning tree problem: Given a graph G and a real number $\alpha \geq 1$, the goal is to find a spanning tree T with minimum $C(T)$ subject to $w(T) \leq \alpha \times w(MST(G))$. For any fixed $\alpha > 1$, an optimal solution can be approximated with a constant ratio in polynomial time.

2. The routing-cost-constrained minimum weight spanning tree problem: Given a graph G and a real number $\alpha \geq 1$, find a spanning tree T with minimum $w(T)$ subject to $C(T) \leq \alpha \times C(MRCT(G))$. For any graph G and a fixed $\alpha > 3/2$, an optimal solution can be approximated with a constant ratio in polynomial time.

3. Let $\alpha_T = C(T)/C(MRCT(G))$ and $\beta_T = w(T)/w(MST(G))$; also let $h(\alpha_T, \beta_T)$ be a specified function for evaluating the total cost of T. Assume that h is monotonically increasing, that is, $h(x+\triangle x, y+\triangle y) \geq h(x,y)$ for all $\triangle x, \triangle y \geq 0$. The goal is to find a spanning tree T with minimum total cost $h(\alpha_T, \beta_T)$. For example, $h(\alpha_T, \beta_T) = \alpha_T \times \beta_T$, or $h(\alpha_T, \beta_T) = x \times \alpha_T + y \times \beta_T$, in which x and y are constants.

By Theorem 6.2, for any real number $\alpha > 1$, a $\left(\frac{k+3}{k+1}\alpha, f(k) + \frac{2}{\alpha-1}\right)$-LART can be constructed in polynomial time, and the ratio

$$\frac{h\left(\frac{k+3}{k+1}\alpha, f(k) + \frac{2}{\alpha-1}\right)}{h(1,1)}$$

is a constant. Since $h(1,1)$ is a trivial lower bound, such a tree is an approximation solution with a constant ratio for any monotonically increasing function h. To obtain a better ratio and reasonable time complexity, suitable α and k should be chosen.

6.5 Summary

In this chapter, we investigate the problems of balancing the building cost and the routing cost of a spanning tree. A LAST is a spanning tree rooted at a source vertex in which the distance from the source to each vertex is no more than a constant times the shortest path length and the total length of the tree edges is no more than a constant times the weight of an MST. Given a graph and a source vertex, an LAST rooted at the source can be constructed in linear time if a minimum spanning tree and a shortest-paths tree are given.

While there is only one source in an LAST, all vertices are considered as sources in an LART. An algorithm for constructing an LART in polynomial time is given. By adjusting the parameters of the algorithm, we can make tradeoffs among the time complexity, the routing cost, and the total edge weight. It is also shown that the result for LART can be extended to some related problems.

Bibliographic Notes and Further Reading

The FIND-LAST algorithm is due to the work of Samir Khuller, Balaji Raghavachari, and Neal E. Young [64]. The algorithm is a modification of the one developed by Baruch Awerbuch, Alan E. Baratz, and David Peleg [5], in which they showed that every graph G has a spanning tree of diameter at most a constant times the diameter of G and of weight at most a constant times the weight of MST. Khuller et al. also showed the bound is tight: For fixed $\alpha > 1$ and $1 \leq \beta < 1 + 2/(\alpha - 1)$, there exists a planar graph containing no (α, β)-LAST. Furthermore, the problem deciding whether a given graph has such a (α, β)-LAST is NP-complete. The FIND-LART algorithm is due to the work of Bang Ye Wu, Kun-Mao Chao, and Chuan Yi Tang [99]. The algorithm is obtained by combining their previous result on MRCT [100] and Khuller's result.

Several results for trees realizing tradeoff between weight and some other distance requirements are available in the literatures. Kadaba Bharath-Kumar and Jeffrey M. Jaffe [13] consider the spanning tree whose weight is α times the one of MST and the sum of the distance from the root to each vertex is at most β times the minimum. They showed that the desired tree exists if $\alpha\beta \geq \Theta(n)$. Another important feature of a network is its diameter. However, finding a minimum diameter subgraph with a budget constraint is NP-hard [78], while a polynomial-time algorithm for finding a minimum diameter spanning tree of a graph with arbitrary edge weights was given in [52].

Considerable work has been done on the *spanner* of general graphs [2, 16, 75] and of Euclidean graphs [22, 71, 89]. In general, a t-spanner of G is a low-weight subgraph of G such that, for any two vertices, the distance within the spanner is at most t times the distance in G. Obviously, the results on spanners such as [2] can be applied to the problems discussed in this chapter. But since the criteria for a spanner are often much stricter, this approach is often less efficient when applied to a weaker distance requirement.

Exercises

6-1. Show that a doubling tree is a Eulerian graph.

6-2. Show that FIND-LAST constructs a $(1 + \sqrt{2}\gamma, 1 + \sqrt{2}/\gamma)$-LAST for any $\gamma > 0$.

6-3. Let T be a minimum spanning tree and $s \in V(T)$ be a specified source. What is the worst ratio $\max_{v \in V}\{d_T(s,v)/d(s,v)\}$? Describe your answer in Big-O notation.

6-4. What is the worst ratio of the weight of a shortest-paths tree to the one of a minimum spanning tree? Describe your answer in Big-O notation.

6-5. Let T be a spanning tree of the metric closure of a graph G. The REMOVE_BAD algorithm in Chapter 4 constructs a spanning tree Y of G with $C(Y) \leq C(T)$. Show that $w(Y) \leq w(T)$ as stated in Section 6.3.4.

6-6. The ADD-PATH subroutine in FIND-LAST is a recursive procedure. Rewrite the subroutine as a nonrecursive procedure.

6-7. Write an algorithm to generate a Eulerian cycle on a doubling tree.

6-8. What happens if we give $\alpha = 1$ in the FIND-LAST algorithm?

6-9. As shown in Lemma 5.9, there is a 2-star which is a 1.577-approximation of a PROCT or an MRCT. Use this result to show how to construct a $\left(1.577\alpha, 2 + \frac{2}{\alpha-1}\right)$-LART. What is the time complexity of your algorithm?

Chapter 7

Steiner Trees and Some Other Problems

In this chapter, we introduce some spanning tree problems not included in the previous chapters. In the first three sections, we discuss the Steiner minimal tree problem, the minimum diameter spanning tree problem, and the maximum leaf spanning tree problem. In the last section, we briefly introduce some other spanning tree problems in both network design and computational biology.

7.1 Steiner Minimal Trees

While a spanning tree spans all vertices of a given graph, a *Steiner tree* spans a given subset of vertices. In the Steiner minimal tree problem, the vertices are divided into two parts: *terminals* and *nonterminal vertices*. The terminals are the given vertices which must be included in the solution. The cost of a Steiner tree is defined as the total edge weight. A Steiner tree may contain some nonterminal vertices to reduce the cost. Let V be a set of vertices. In general, we are given a set $L \subset V$ of terminals and a metric defining the distance between any two vertices in V. The objective is to find a connected subgraph spanning all the terminals of minimal total cost. Since the distances are all nonnegative in a metric, the solution is a tree structure. Depending on the given metric, two versions of the Steiner tree problem have been studied.

- **(Graph) Steiner minimal trees (SMT)**: In this version, the vertex set and metric is given by a finite graph.

- **Euclidean Steiner minimal trees (Euclidean SMT)**: In this version, V is the entire Euclidean space and thus infinite. Usually the metric is given by the Euclidean distance (L^2-norm). That is, for two points with coordinates (x_1, y_2) and (x_2, y_2), the distance is

$$\sqrt{(x_1 - x_2)^2 + (y_1 - y_2)^2}.$$

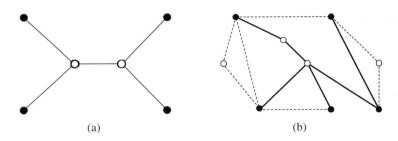

FIGURE 7.1: Examples of Steiner minimal trees. (a) A Euclidean Steiner minimal tree; and (b) a graph Steiner minimal tree. The black points are terminals and the white points are nonterminals.

In some applications such as VLSI routing, L^1-norm, also known as *rectilinear distance*, is used, in which the distance is defined as

$$|x_1 - x_2| + |y_1 - y_2|.$$

Figure 7.1 illustrates a Euclidean Steiner minimal tree and a graph Steiner minimal tree.

7.1.1 Approximation by MST

Let $G = (V, E, w)$ be an undirected graph with nonnegative edge weights. Given a set $L \subset V$ of terminals, a Steiner minimal tree is the tree $T \subset G$ of minimum total edge weight such that T includes all vertices in L.

PROBLEM: Graph Steiner Minimal Tree
INSTANCE: A graph $G = (V, E, w)$ and a set $L \subset V$ of terminals
GOAL: Find a tree T with $L \subset V(T)$ so as to minimize $w(T)$.

The decision version of the SMT problem was shown to be NP-complete by a transformation from the EXACT COVER BY 3-SETS problem [63]. We shall focus on some approximation results.

A well-known method to approximate an SMT is to use a minimal spanning tree (MST). First we construct the metric closure on L, i.e., a complete graph with vertices L and edge weights equal to the shortest path lengths. Then we find an MST on the closure, in which each edge corresponds to one shortest path on the original graph. Finally the MST is transformed back to a Steiner tree by replacing each edge with the shortest path and some straightforward postprocessing to remove any possible cycle.

Algorithm: MST-STEINER
Input: A graph $G = (V, E, w)$ and a terminal set $L \subset V$.
Output: A Steiner tree T.
1:     Construct the metric closure G_L on the terminal set L.
2:     Find an MST T_L on G_L.
3:     $T \leftarrow \emptyset$.
4:     **for** each edge $e = (u, v) \in E(T_L)$ **do**
4.1:   Find a shortest path P from u to v on G.
4.2:   **if** P contains less than two vertices in T **then**
                Add P to T;
        else
            Let p_i and p_j be the first and the last vertices already in T;
            Add subpaths from u to p_i and from p_j to v to T.
5:     Output T.

Basically we replace each edge in T_L with the corresponding shortest path at Step 4. But if there are two vertices already in the tree, adding the path will result in cycles. In this case we only insert the subpaths from the terminals to the vertices already in the tree. It avoids any cycle and ensures that the terminals are included. As a result, we can see that the algorithm returns a Steiner tree.

Example 7.1
Figure 7.2 shows a sample execution of the algorithm. (a) is the given graph G, in which $L = \{v_i | 1 \leq i \leq 5\}$ is the terminal set and u_i, $1 \leq i \leq 4$ are nonterminal vertices. (b) and (c) are the metric closure G_L and a minimum spanning tree T_L on G_L respectively. Initially T is empty. Suppose that edge $(v_1, v_4) \in E(T_L)$ is first chosen edge. The corresponding shortest path in G is (v_1, u_1, u_2, v_4). Since all vertices on the path are not in T, the whole path is inserted into T (Frame(d)). At the second iteration, edge $(v_2, v_3) \in E(T_L)$ is chosen. The corresponding shortest path is (v_2, u_1, u_2, v_3). However, u_1 and u_2 are already in T. Therefore, only (v_2, u_1) and (u_2, v_3) are inserted (Frame (e)). At the third iteration, edge (v_2, v_5) is chosen, and edge (v_2, v_5) is inserted (Frame (f)). Finally, the last edge in T_L is (v_1, v_2). Since both v_1 and v_2 are already in T, no edge is inserted, and the tree in Frame (f) is the output tree. □

We are interested in how good the output tree is. Let $G_L = (L, E(G_L), \bar{w})$. First we observe that $w(T) \leq \bar{w}(T_L)$ since at most all the shortest paths are inserted. We want to compare T_L with a Steiner minimal tree. Let $smt(G, L)$ be the Steiner minimal tree. Consider a Eulerian tour X on the doubling tree (see Figure 6.2 for the definition of a Eulerian tour on a doubling tree). Since the tour traverses each edge exactly twice, $w(X) = 2w(smt(G, L))$. On the

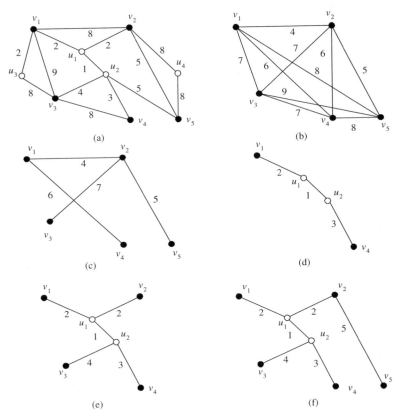

FIGURE 7.2: A sample execution of Algorithm MST-STEINER.

other hand, since the tour visits all the terminals, we have

$$w(X) \geq w(\text{tsp}(G_L)) \geq w(T_L),$$

in which $\text{tsp}(G_L)$ is a minimal Hamiltonian cycle on G_L— an optimal solution of the *Traveling Salesperson problem*. Consequently we have

$$w(T) \leq w(T_L) \leq 2w(\text{smt}(G, L)).$$

To see the bound is asymptotically tight, we show an extreme example in Figure 7.3. Suppose that there are k terminals and one nonterminal vertex. The terminals form a cycle with $w(e) = 2$ for each edge e in the cycle. The nonterminal vertex is connected to each terminal with an edge of unit length. The Steiner minimal tree is the star centered at the nonterminal vertex and

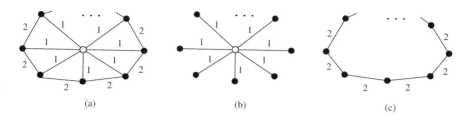

FIGURE 7.3: (a) A graph in which the white vertex is a nonterminal vertex and others are terminals. (b) A Steiner minimal tree. (c) A minimum spanning tree of the subgraph induced on the terminals.

has weight k. Meanwhile an MST is the path consisting of the terminals and has weight $2(k-1)$. Therefore the ratio is $(2-2/k)$.

The time complexity of the algorithm is $O(|V||L|^2)$, dominated by the construction of the metric closure. In summary we have the following theorem:

THEOREM 7.1
The MST-STEINER *algorithm finds a 2-approximation of an SMT of a general graph in $O(|V||L|^2)$ time, where V is the vertex set and L is the terminal set. Furthermore, the ratio is asymptotically tight.*

Define the *Steiner ratio* as

$$\rho = \frac{w(\mathrm{mst}(G_L))}{w(\mathrm{smt}(G,L))}.$$

The above theorem shows that the Steiner ratio in general metric is 2. Since the Euclidean space is a restricted version of the general metric, it is not surprising that the Euclidean Steiner ratio is smaller than that in general metric. In fact, it was shown that the ratio is $2/\sqrt{3}$. But the proof is beyond the scope of this book (see the Bibliographic Notes).

7.1.2 Improved approximation algorithms

The MST-STEINER algorithm approximates an SMT without any help from the nonterminal vertices. In the following we introduce an approximation algorithm proposed by Alexander Zelikovsky [105], which uses the Steiner vertices to obtain a better performance guarantee.

Let F be a metric graph. By $F \setminus e$, we denote the resulting graph by *contracting* edge e, i.e., reducing its length to 0. A *triple* is a 3-tuple (u_1, u_2, u_3) of terminals. For a triple $z = (u_1, u_2, u_3)$, define $F \setminus z$ as the resulting graph by contracting edges (u_1, u_2) and (u_2, u_3). In other words, $F \setminus z$ is obtained by contracting the three terminals to one point. Suppose that x is a Steiner vertex and $\delta(x) = w(x, u_1) + w(x, u_2) + w(x, u_3)$ is the total distance from x to the three terminals. We may have a Steiner tree by combining an MST of the contracted graph $F \setminus z$ with the star centering at x and spanning the three terminals. The cost is given by $w(\mathrm{mst}(G \setminus z)) + \delta(x)$. Let

$$\mathrm{gain}(z) = w(\mathrm{mst}(F)) - (w(\mathrm{mst}(F \setminus z)) + \delta(x)).$$

If $gain(z) > 0$, clearly we may have a better Steiner tree. The algorithm repeatedly contracts the triple with greatest gain until no triple with positive gain exists. The resulting Steiner tree is given by an MST on the terminals and the Steiner vertices used to contract triples. We now list the algorithm in detail.

Algorithm: ZELIKOVSKY-STEINER
Input: A graph $G = (V, E, w)$ and a terminal set $L \subset V$.
Output: A Steiner tree T.
1:   $F \leftarrow G_L$; $W \leftarrow \emptyset$;
2:   **for** each triple z **do**
        find a vertex z_{min} minimizing $\sum_{v \in z} w(v, z_{min})$;
        let $\delta(z) \leftarrow \sum_{v \in z} w(v, z_{min})$;
3:   **repeat** forever
        find the triple z maximizing
           $\mathrm{gain}(z) = w(\mathrm{mst}(F)) - w(\mathrm{mst}(F \setminus z)) - \delta(z)$;
        if $\mathrm{gain}(z) \leq 0$ **then** go to Step 4;
        $F \leftarrow F \setminus z$;
        insert z_{min} into W;
4:   Find $T = \mathrm{mst}(G_{L \cup W})$ and output T.

Example 7.2 illustrates how the algorithm runs.

Example 7.2
Figure 7.4(a) is the input graph G, in which $L = \{v_1, v_2, v_3, v_4\}$ is the terminal set, and u_1 and u_2 are nonterminal vertices. (b) is the metric closure G_L. We can see that the total edge weight of the minimum spanning tree is 12. In this example, there are only two triples $z_1 = (v_1, v_2, v_3)$ and $z_2 = (v_2, v_3, v_4)$ with finite δ values. For any other triple, for example (v_1, v_2, v_4), there is no Steiner vertex adjacent to all the three vertices in the triple. Therefore $\delta(v_1, v_2, v_4) = \infty$, and we can ignore it. For z_1, we have $z_{min} = u_1$ and

$$\delta(z_1) = w(u_1, v_1) + w(u_1, v_2) + w(u_1, v_3) = 9.$$

Similarly, for z_2, $z_{min} = u_2$ and $\delta(z-2) = 6$. The resulting graphs by contracting z_1 and z_2 are shown in (c) and (d) respectively. We have

$$\mathrm{gain}(z_1) = w(\mathrm{mst}(F)) - w(\mathrm{mst}(F \setminus z_1)) - \delta(z_1) = 12 - 3 - 9 = 0,$$

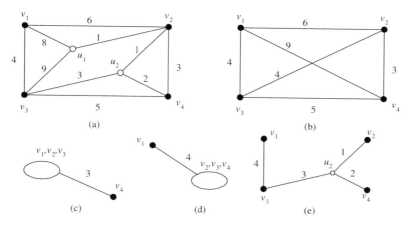

FIGURE 7.4: A sample execution of the ZELIKOVSKY-STEINER algorithm.

and

$$\text{gain}(z_2) = w(\text{mst}(F)) - w(\text{mst}(F \setminus z_2)) - \delta(z_2) = 12 - 4 - 6 = 2.$$

Since $\text{gain}(z_2) > \text{gain}(z_1)$, we choose z_2, and u_2 is inserted into W. After z_2 is contracted, the resulting graph contains only two vertices and the algorithm goes to Step 4. Finally a minimum spanning tree of the metric closure on vertex set $L \cup \{u_2\}$ is constructed and output (Frame (e)). □

The following theorem comes from [105].

THEOREM 7.2
For a graph $G = (V, E, w)$ and a set L of terminals, a Steiner minimal tree can be approximated with ratio 11/6 in $O(|V||E| + |L|^4)$ time.

Instead of giving the details, we outline the Zelikovsky's proof. The approximation ratio in the theorem is shown in two parts. First it is shown that there exists a set of triples such that a Steiner tree obtained by the contraction of the triples is of cost within a factor 5/3 of the optimal, i.e.,

$$w(T_1) \leq \frac{5}{3} w(\text{smt}(G, L)), \tag{7.1}$$

in which T_1 is the Steiner tree with optimal contraction on triples. Second, the total gain of a greedy contracting sequence as in the algorithm is at least one half of that of any set of triples, i.e.,

$$w(\text{mst}(G_L)) - w(T_2) \geq \frac{1}{2} \left(w(\text{mst}(G_L)) - w(T_1) \right), \tag{7.2}$$

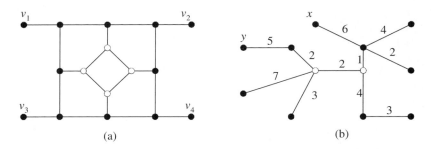

FIGURE 7.5: Diameters, centers and radii of (a) an unweighted graph; and (b) a weighted tree.

in which T_2 is the tree obtained by a greedy contracting sequence. By (7.2), we have
$$w(T_2) \le \frac{1}{2}\left(w(\mathrm{mst}(G_L)) + w(T_1)\right).$$
Since $w(\mathrm{mst}(G_L)) \le 2w(\mathrm{smt}(G,L))$ and $w(T_1) \le \frac{5}{3}w(\mathrm{smt}(G,L))$ by (7.1), we obtain
$$w(T_2) \le \frac{1}{2}\left(2w(\mathrm{smt}(G,L)) + \frac{5}{3}w(\mathrm{smt}(G,L))\right) \le \frac{11}{6}w(\mathrm{smt}(G,L)).$$

7.2 Trees and Diameters

7.2.1 Eccentricities, diameters, and radii

Let $G = (V, E, w)$ be a graph and $U \subset V$. By $D_G(v, U)$, we denote the maximum distance from vertex v to any vertex in U. For a vertex v, the *eccentricity* of v is the maximum of the distance to any vertex in the graph, i.e., $\max_{u \in V}\{d_G(v, u)\}$ or $D_G(v, V)$. The *diameter* of a graph is the maximum of the eccentricity of any vertex in the graph. (The term "diameter" is overloaded. It is defined as the maximum eccentricity and also as the path of length equal to the maximum eccentricity.) In other words, the diameter is the longest distance between any two vertices in the graph. Recall that the distance between two vertices is the length of their shortest path in the graph. It should not be confused with the longest path in the graph.

The *radius* of a graph is the minimum eccentricity among all vertices in the graph, and a *center* of a graph is a vertex with eccentricity equal to the radius. For a general graph, there may be several centers and a center is not necessarily on a diameter. For example, in Figure 7.5(a), the shortest path between v_1 and v_4 is a diameter of length 6. Meanwhile v_2 and v_3 are endpoints of another diameter. The four vertices represented by white circles are centers of the graph, and the radius is 4. Note that the centers are not on any diameter. The diameter, radius and center of a graph can be found by computing the distances between all pairs of vertices. It takes $O(|V||E| + |V|^2 \log |V|)$ time for a general graph.

Any pair of vertices has a unique simple path in a tree. For this reason, the diameter, radius and centers of a tree are more related. For an unweighted tree $T = (V, E)$, it can be easily verified that

$$2 \times \text{radius} - 1 \leq \text{diameter} \leq 2 \times \text{radius}. \tag{7.3}$$

For positive weighted tree $T = (V, E, w)$, we can also have

$$2 \times \text{radius} - \max_e \{w(e)\} \leq \text{diameter} \leq 2 \times \text{radius}. \tag{7.4}$$

Figure 7.5(b) illustrates an example of the diameter, radius, and center of a tree. The two vertices represented by white circles are the centers of the tree. There are four diameters of length 16 in the tree. Vertices x and y are the endpoints of a diameter. The radius is 9, and the centers are on the diameters.

More efficient algorithms for computing diameter, radius, and centers are available if the graph is a tree. Let $T = (V, E, w)$ be a rooted tree. By T_r, we denote the subtree rooted at vertex $r \in V$, which is the subgraph induced on vertex r and all its descendants. Let child(r) denote the set of children of v. The eccentricity of the root of a tree can be computed by the following recurrence relation.

$$D_{T_r}(r, V(T_r)) = \max_{s \in child(r)} \{D_{T_s}(s, V(T_s)) + w(r, s)\}. \tag{7.5}$$

By a recursive algorithm or an algorithm visiting the vertices in postorder, the eccentricity can be computed in linear time since each vertex is visited once. To make it clear, we give a recursive algorithm in the following.

Algorithm: ECCENT(T_r)
Input: A tree $T_r = (V, E, w)$ rooted at r.
Output: The eccentricity of r in T_r.
1:    if r is a leaf then
         return 0;
2:    for each child s of r do
         compute ECCENT(T_s) recursively;
3:    return $\max_{s \in child(r)}\{\text{ECCENT}(s) + w(r, s)\}$.

To find the eccentricity of a vertex in an unrooted tree, we can root the tree at the vertex and employ the ECCENT algorithm. Therefore, we have the next lemma.

LEMMA 7.1
The eccentricity of a vertex in a tree can be computed in linear time.

Let x, y and z be three vertices in a tree T. It can be easily verified that the three paths $SP_T(x,y)$, $SP_T(x,z)$ and $SP_T(y,z)$ intersect at a vertex. Define $c(x,y,z)$ to be the intersection vertex. We have

$$d_T(x, c(x,y,z)) = \frac{1}{2}(d_T(x,y) + d_T(x,z) - d_T(y,z)), \qquad (7.6)$$

or equivalently

$$d_T(x, c(x,y,z)) = \frac{1}{2}(d_T(x,y) + d_T(x,z) + d_T(y,z)) - d_T(y,z). \qquad (7.7)$$

We now derive some properties to help us find the diameter of a tree.

FACT 7.1
Suppose that $SP_T(v_1, v_2)$ is a diameter of T and r is a vertex on the diameter. For any vertex x, $d_T(x,r) \leq \max\{d_T(r,v_1), d_T(r,v_2)\}$.

PROOF Otherwise $SP_T(x, v_1)$ or $SP_T(x, v_2)$ is a path longer than the diameter. ☐

The property can be easily extended to the case where r is not on the diameter. Let $u = c(r, v_1, v_2)$. Without loss of generality, let $d_T(u, v_1) \geq d_T(u, v_2)$. For any vertex x, we have $d_T(x, u) \leq d_T(v_1, u)$ by Fact 7.1. Then

$$d_T(x, r) \leq d_T(x, u) + d_T(u, r) \leq d_T(v_1, u) + d_T(u, r) = d_T(v_1, r),$$

which implies that v_1 is the farthest vertex to r. We can conclude that, for any vertex, one of the endpoints of a diameter must be the farthest vertex. Furthermore the converse of the property is also true.

Let r be any vertex in a tree and v_3 be the vertex farthest to r. We shall show that v_3 must be an endpoint of a diameter. Suppose that $SP_T(v_1, v_2)$ is a diameter. By the above property, one of the endpoints, say v_1, is the farthest vertex to r, i.e.,

$$d_T(r, v_1) = d_T(r, v_3) \geq d_T(r, v_2).$$

Let $u_1 = c(r, v_1, v_2)$ and $u_2 = c(r, v_1, v_3)$. Then u_2 must be on the path $SP_T(v_1, u_1)$, for otherwise

$$\begin{aligned}d_T(v_1, v_3) &= d_T(v_1, u_2) + d_T(u_2, v_3)\\&= 2d_T(v_1, u_2) > 2d_T(v_1, u_1)\\&> d_T(v_1, v_2),\end{aligned}$$

a contradiction. As a result, $d_T(v_3, v_2) = d_T(v_1, v_2)$ and $SP_T(v_3, v_2)$ is also a diameter. We have the next lemma.

LEMMA 7.2
Let r be any vertex in a tree T. If v is the farthest vertex to r, the eccentricity of v is the diameter of T.

The following algorithm uses the property to find the diameter of a tree.

Algorithm: TREEDIAMETER
Input: A tree $T = (V, E, w)$.
Output: The diameter of T.
1: Root T at an arbitrary vertex r.
2: Use ECCENT to find the farthest vertex v to r.
3: Root T at v.
4: Use ECCENT to find the eccentricity of v.
5: Output the eccentricity of v as the diameter of T.

It is obvious that the algorithm runs in linear time. The radius and the center can be obtained from a diameter. Suppose that $P = SP_T(v_1, v_2)$ is a diameter. Starting at v_1 and traveling along the path P, we compute the distance $d_T(u, v_1)$ for each vertex u on the path. Let u_1 be the last encountered vertex such that $d_T(v_1, u_1) \leq \frac{1}{2}w(P)$ and u_2 be the next vertex to u_1 (Figure 7.6(a)). By the definition of u_1, u_2 is the first encountered vertex such that $d_T(v_1, u_2) > \frac{1}{2}w(P)$. We claim that u_1 or u_2 is a center of the tree. Let $P_1 = SP_T(v_1, u_1)$ and $P_2 = SP_T(u_2, v_2)$. First, by Fact 7.1, the eccentricities of u_1 and u_2 must be $d_T(u_1, v_2)$ and $d_T(u_2, v_1)$ respectively. Otherwise P cannot be a diameter. For any vertex x connected to P at a vertex in P_1, we have $d_T(x, v_2) > d_T(u_1, v_2)$. Similarly, for any vertex x connected to P at a vertex in P_2, we have $d_T(x, v_1) > d_T(u_2, v_1)$. Consequently the eccentricity of any vertex is at least $\min\{d_T(u_1, v_2), d_T(u_2, v_1)\}$, and u_1 or u_2 must be a center. Therefore the center as well as the radius of a tree can be computed in linear time since the diameter can be found in linear time.

THEOREM 7.3
The diameter, radius, and center of a tree can be computed in linear time.

7.2.2 The minimum diameter spanning trees

For a graph $G = (V, E, w)$, the *minimum diameter spanning tree* (MDST) of G is a spanning tree of minimum diameter among all possible spanning trees. Removing edges from a graph usually increases the diameter of the graph. In general, we cannot expect to find a spanning tree with the same diameter as the underlying graph. For example, consider an unweighted complete graph.

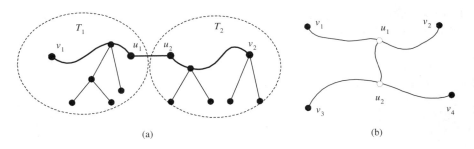

FIGURE 7.6: (a) Finding a center on a diameter. (b) Two diameters cannot be disjoint.

The diameter of the graph is one, while the minimum diameter of any spanning tree is two. We shall introduce an efficient algorithm for finding an MDST of a given graph. First we examine some properties of the diameters of a tree. We assume that all edge lengths are positive.

FACT 7.2
Two diameters of a tree cannot be disjoint.

PROOF Suppose that $SP_T(v_1, v_2)$ and $SP_T(v_3, v_4)$ are two disjoint diameters of a tree T. Let $u_1 = c(v_3, v_1, v_2)$ and $u_2 = c(v_1, v_3, v_4)$ (Figure 7.6(b)). We have

$$d_T(v_1, v_3) + d_T(v_2, v_4) = d_T(v_1, v_2) + d_T(v_3, v_4) + 2d_T(u_1, u_2)$$
$$> 2d_T(v_1, v_2)$$

since $d_T(v_1, v_2) = d_T(v_3, v_4)$ is the diameter and $d_T(u_1, u_2) > 0$. It implies that the path from v_1 to v_3 or the path from v_2 to v_4 is longer than the diameter, a contradiction. □

Let $\mathcal{P}$ be a set of more than two paths of a tree and the paths intersect each other. One can easily verify that all the paths in $\mathcal{P}$ share a common vertex. Otherwise there exists a cycle, which contradicts the definition of a tree. Therefore we can have the next property.

FACT 7.3
All diameters of a tree share at least one common vertex.

Let v_1 and v_2 be the endpoints of a diameter of a tree T (Figure 7.6(a)). There exists an edge $(u_1, u_2) \in E(T)$ such that the removal of the edge results in two subpaths at most one half long as the diameter. Let T_1 and T_2 be the two subgraphs obtained by removing (u_1, u_2) from T, in which u_1 and v_1 are in T_1. As mentioned early, u_1 or u_2 (or both) is a center of T. Without loss of generality, we assume $d_T(v_1, u_1) \leq d_T(v_2, u_2)$. It implies that u_1 is a center and the radius of the tree is $d_T(u_1, v_2)$.

Now suppose that T is a MDST of a graph $G = (V, E, w)$. We shall derive some properties helpful for computing an MDST of a graph. The discussion is divided into two cases:

- The radius is exactly one half of the diameter, i.e., $d_T(u_1, v_2) = d_T(u_1, v_1)$.

- $d_T(u_1, v_2) \neq d_T(u_1, v_1)$.

For the first case, let's consider a shortest-paths tree Y rooted at u_1. Since, for any vertex v,

$$d_Y(u_1, v) = d(u_1, v) \leq d_T(u_1, v_1),$$

the eccentricity of u_1 in Y is the same as in T, and Y is also an MDST of G.

LEMMA 7.3
If the diameter is exactly two times the radius of an MDST of a graph G, any shortest-paths tree rooted at a center of an MDST is also an MDST.

Now we turn to the second case. To simplify the discussion, we start at an easier case where G is a metric graph, e.g., any graph in the Euclidean space. For all vertices in $V(T_1)$, we construct a star centered at u_1. Similarly construct a star centered at u_2 for all vertices in $V(T_2)$. Let X be the tree containing the two stars and the edge (u_1, u_2). Obviously the diameter of X is the same as the diameter of T, and X is also a MDST of G. We have the next lemma. Note that it covers a metric version of Lemma 7.3.

LEMMA 7.4
For a metric graph, there exists an MDST with an edge (u_1, u_2) such that, for any other vertex v, either (u_1, v) or (u_2, v) is in the tree.

It should be noted that simply connecting all vertices to u_1 does not always give us an MDST. It is possible that there exist two vertices v_3 and v_4 in T_2 such that $d(v_3, u_1) + d(v_4, u_1)$ is larger than the diameter of T. Now we consider the case where G is a general graph. In fact, we have a similar result for general graphs. The difficulty of showing the property is that a pair of vertices may not be adjacent and an edge may be longer than the shortest path between the two endpoints.

We bipartition the vertex set V into V_1 and V_2, in which

$$V_1 = \{v | d(v, u_1) - d(v, u_2) \leq d(v_1, u_1) - d(v_2, u_2)\}.$$

For any vertex $v \in V_1$, we claim that the shortest path from v to u_1 contains only vertices in V_1. Let $u \in V(SP_G(v, u_1))$. If $u \in V_2$, we have

$$d(u, u_1) - d(u, u_2) > d(v_1, u_1) - d(v_2, u_2),$$
$$(d(v, u) + d(u, u_1)) - (d(v, u) + d(u, u_2)) > d(v_1, u_1) - d(v_2, u_2),$$
$$d(v, u_1) - d(v, u_2) > d(v_1, u_1) - d(v_2, u_2)$$

since $d(v, u_2) \le d(v, u) + d(u, u_2)$. Similarly, for any vertex in V_2, the shortest path to u_2 contains only vertices in V_2. Thus, there exists a spanning tree X in which the path from u_1 (and u_2) to any vertex in V_1 (and V_2, respectively) is a shortest path. It is obvious that the diameter of X is the same as the diameter of T.

LEMMA 7.5
Let $G = (V, E, w)$ be a graph. There exists a minimum diameter spanning tree T with an edge $(u_1, u_2) \in E(T)$ such that $d_T(v, u_1) = d(v, u_1)$ or $d_T(v, u_2) = d(v, u_2)$ for any vertex $v \in V$.

7.2.2.1 An algorithm

The property in Lemma 7.5 provides us a method to construct an MDST of a given graph. Recall that $D_G(v, U)$ denotes $\max_{u \in U} d_G(v, u)$ for a vertex v and a vertex set U in a graph G. For each edge (u_1, u_2), we find a bipartition (V_1, V_2) of the vertex set such that $D_G(u_1, V_1) + D_G(u_2, V_2)$ is minimized subject to

$$d(v_1, u_1) - d(v_1, u_2) \le d(v_2, u_1) - d(v_2, u_2)$$

for any vertex $v_1 \in V_1$ and $v_2 \in V_2$. The diameter of the tree corresponding to edge (u_1, u_2) is $w(u_1, u_2) + D_G(u_1, V_1) + D_G(u_2, V_2)$, and an MDST can be found by trying all the edges and choosing the best.

Algorithm: MDST
Input: A graph $G = (V, E, w)$.
Output: A minimum diameter spanning tree T.
1: Compute the shortest path length for each pair of vertices.
2:    $\delta \leftarrow \infty$.
3:    **for** each edge $(u_1, u_2) \in E$ **do**
4:       $\alpha(v) \leftarrow d(v, u_1) - d(v, u_2)$ for each $v \in V$.
5:       Sort and relabel the vertices such that $\alpha(v_i) \le \alpha(v_{i+1})$.
6:       **for** each i, $\alpha(v_i) < \alpha(v_{i+1})$, **do**
7:          $V_1 = \{v_j | j \le i\}$; $V_2 = \{v_j | j > i\}$;
8:          **if** $D_G(u_1, V_1) + D_G(u_2, V_2) + w(u_1, u_2) < \delta$
9:             $\delta \leftarrow D_G(u_1, V_1) + D_G(u_2, V_2) + w(u_1, u_2)$;
10:            $\alpha \leftarrow \alpha(v_i); (x, y) \leftarrow (u_1, u_2)$;
11: Reconstruct V_1 and V_2 from (x, y) and α.

12: Find a shortest-paths tree T_1 rooted at x and spanning V_1.
13: Find a shortest-paths tree T_2 rooted at y and spanning V_2.
14: Output $T = T_1 \cup T_2 \cup \{(x, y)\}$.

The correctness of the algorithm follows Lemma 7.5 and the above discussion. Step 1 takes $O(|V||E| + |V|^2 \log |V|)$ time for computing the all-pair shortest path lengths. At Step 5, sorting $\alpha(v)$ takes $O(|V| \log |V|)$ time for each edge and thus $O(|E||V| \log |V|)$ time in total. At Step 8, we need to find the maximum distances from any vertex in V_1 and V_2 to u_1 and u_2 respectively. Since the vertices are moved from V_2 to V_1 incrementally, by using a data structure, such as a heap, to store vertices in V_2, each maximum distance can be found in $O(\log |V|)$ time. Therefore the time complexity of the step is $O(|E||V| \log |V|)$ in total. Finally the two shortest-paths trees can be constructed in $O(|E| + |V| \log |V|)$ time. In summary, the total time complexity of the algorithm is $O(|E||V| \log |V|)$.

THEOREM 7.4
Given a graph $G = (V, E, w)$, a minimum diameter spanning tree of G can be computed in $O(|E||V| \log |V|)$ time.

7.2.2.2 The absolute 1-center

As mentioned in the previous paragraphs, the center of a graph is the *vertex* whose eccentricity is equal to the radius. The *absolute 1-center* has a similar meaning but may be any *point* in the graph. Different from the graph model usually used, any interior point in addition to the two endpoints (vertices) of an edge is referred to as a point in the graph. Let $G = (V, E, w)$ and $A(G)$ denote the set of points on the edges of G. The edge lengths induce a distance function on $A(G)$. We overload the notation $d(x, y)$ for the distance between two points x and y in $A(G)$. The absolute 1-center of G is the point x minimizing the function

$$f(x) = \max_{v \in V} d(x, v).$$

For a general graph, there may be several absolute 1-centers. However, for a tree, the absolute 1-center is unique. The reason is that all diameters of a tree can not be disjoint (Fact 7.2). Furthermore, the diameter of the tree is exactly two times the eccentricity of the absolute 1-center. An absolute 1-center of a graph defines a minimum diameter spanning tree.

THEOREM 7.5
Let x be an absolute 1-center of a graph $G = (V, E, w)$. Any shortest-paths tree rooted at x is a minimum diameter spanning tree of G.

PROOF Let $\delta(H)$ denote the diameter of graph H. Let T be any spanning tree of G and u is the absolute 1-center of T. We have

$$\delta(T) = 2D_T(u, V).$$

Assume that Y is any shortest-paths tree rooted at x, an absolute 1-center of G. It follows that x is the midpoint of every diameter of Y. Then,

$$\begin{aligned}\delta(Y) &= 2D_Y(x, V) \\ &= 2D_G(x, V) \\ &\leq 2D_G(u, V) \\ &\leq 2D_T(u, V) \\ &= \delta(T).\end{aligned}$$

That is, the diameter of Y is less than or equal to the diameter of any spanning tree. We conclude any shortest-paths tree rooted at an absolute 1-center of a graph is a minimum diameter spanning tree. □

The most efficient algorithm for finding an absolute 1-center of a graph takes $O(|V||E| + |V|^2 \log |V|)$ time [62]. We have the following corollary.

COROLLARY 7.1
Given a graph $G = (V, E, w)$, a minimum diameter spanning tree of G can be found in $O(|V||E| + |V|^2 \log |V|)$ time.

7.3 Maximum Leaf Spanning Trees

A leaf in a tree is a vertex with degree one, i.e., incident with only one edge. For a connected undirected graph G, a *maximum leaf spanning tree* (MLST) of G is a spanning tree maximizing the number of leaves. Given a graph G and an integer B, the problem of asking if there is a spanning tree of G with B or more leaves has been shown to be NP-complete. Also the MLST problem was shown to be MAX SNP-hard, and therefore it is impossible to derive a PTAS unless NP=P. In this section, we shall introduce a 3-approximation algorithm due to the work of Lu and Ravi [73]. First we introduce some notations.

7.3.1 Leafy trees and leafy forests

For a graph H, let $V_i(H) \subset V(H)$ denote the set of vertices with degree i in H for any $i \geq 0$. Also let $\overline{V}_i(H) = \bigcup_{j \geq i} V_j(H)$ be the set of vertices whose degrees are at least i in H. Let T be a tree. The leaves of T are the vertices

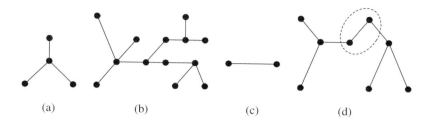

FIGURE 7.7: (a) and (b) are leafy trees, while (c) and (d) are not leafy trees. The tree in (c) has no vertex of degree more than two, and the tree in (d) has vertices of degree two but adjacent to another vertex of degree two.

in $V_1(T)$. The vertices in $\overline{V}_3(T)$ can be thought of as the branching vertices since each of them creates at least one more branch in T. A tree with only one edge has two leaves and no branch vertex. Inserting a branching vertex into the tree, at least one leaf is also needed to insert. Therefore the reader can easily verify the following equation by induction.

$$|\overline{V}_3(T)| \leq |V_1(T)| - 2. \tag{7.8}$$

DEFINITION 7.1 A tree T is leafy if $\overline{V}_3(T) \neq \emptyset$ and each vertex in $V_2(T)$ is adjacent to two vertices in $\overline{V}_3(T)$.

The intuition of the definition of a leafy tree is that a tree with only few vertices of degree two may have many leaves. Figure 7.7 gives some examples. Precisely speaking, we shall establish a lower bound of the number of leaves in a leafy tree. Let T be a leafy tree. Since each vertex in $V_2(T)$ is adjacent to two branching vertices, we have

$$|V_2(T)| \leq |\overline{V}_3(T)| - 1.$$

By Equation (7.8), we have

$$\begin{aligned}|V(T)| &= |V_1(T)| + |V_2(T)| + |\overline{V}_3(T)| \\ &\leq |V_1(T)| + 2|\overline{V}_3(T)| - 1 \\ &\leq 3|V_1(T)| - 5.\end{aligned}$$

We state the lower bound as a lemma.

LEMMA 7.6
Let T be a leafy tree. Then $|V(T)| \leq 3|V_1(T)| - 5$.

If a graph G has a leafy spanning tree T, clearly T is a 3-approximation of the maximum leaf spanning tree of G since any spanning tree has at most

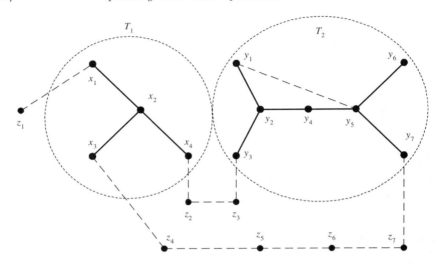

FIGURE 7.8: A maximal leafy forest.

$|V(G)| - 1$ leaves and the number of leaves in a leafy spanning tree is at least one-third of the optimal asymptotically. Unfortunately, a graph may have no leafy spanning tree. We define the *leafy forest*.

DEFINITION 7.2 *A forest F of a graph G is a leafy forest if all its components are leafy trees. A leafy forest is maximal if it is not contained in any other leafy forest.*

It is not necessary that a leafy forest spans all vertices of the graph. What we need is a maximal leafy forest. Let F be a maximal leafy forest consisting of leafy trees T_i for $1 \leq i \leq k$. By the maximality, we can obtain the following properties. We shall use Figure 7.8 to illustrate the properties. In the figure, the bold lines represent the edges in the forest and the dash lines represent edges in the graph but not in the forest.

FACT 7.4
If $v \in \overline{V}_2(T_i)$, for any $u \notin V(T_i)$, we have $(u,v) \notin E(G)$.

Otherwise edge (u,v) can be added to F, which contradicts the maximality of F. Vertex x_2 is an example. If $(x_2, y_1) \in E(G)$, the two trees can be merged without conflicting the definition of a leafy tree. Similarly if $(x_2, z_2) \in E(G)$, we can include z_2 into the tree.

FACT 7.5

Let v be a vertex in T_i and both u_1 and u_2 be vertices adjacent to v in G. If u_1 is not in F, u_2 must be in T_i.

Otherwise both the two neighbors could be added to T_i. The case that v has one neighbor not in F happens when v is a leaf in some leafy tree. Inserting its neighbor into the tree conflicts with the requirement of a leafy tree. Vertex x_1 is an example. It has two neighbors x_2 and z_1. Since z_1 is not in F, x_2 must be in the same tree as x_1. If x_1 had another neighbor, say y_1, not in T_1, we could include both the edges (x_1, z_1) and (x_1, y_1) into the forest. In such a case, x_1 is a branching vertex and it would be a valid leafy tree.

FACT 7.6

If vertex v has two neighbors not in F, the degree of v in G is two.

In this case, vertex v is not in F. If v had any other neighbor, v and all its neighbors could be added to F since they form a star centered at v. Vertex z_5 is an example. It has two neighbors z_4 and z_6 not in F. If z_5 was adjacent to a vertex, say z_2, not in F, z_5 and all its neighbors formed a leafy tree. If z_5 was adjacent to a vertex, say y_7, in F, z_5 and all its neighbors could be included into the forest since z_5 would be a branching vertex in such a case.

7.3.2 The algorithm

The algorithm computes an approximation of an MLST in two steps. First it constructs a maximal leafy forest. The forest consists of several leafy trees and there may be some vertices not in the forest, which are viewed as trees of singleton. In the second step, the leafy trees and the trees of singleton are combined into a spanning tree by adding edges across the trees. The following is the algorithm for constructing a maximal leafy forest.

Algorithm: LEAFY-FOREST
Input: A graph $G = (V, E, w)$.
Output: A maximal leafy forest F.
1:     Make each vertex v a tree of singleton and $deg(v) \leftarrow 0$;
2:     **for** each vertex v **do**
3:         Find the set S of trees containing a vertex u adjacent to v;
4:         **if** $deg(v) + |S| \geq 3$ **then**
5:             Union the trees in S and the tree containing v;
6:             Update the degrees of v and the adjacent vertices in S;
7:     Let F be the union of the nonsingleton trees;
8:     Output F as a maximal leafy forest.

Initially each vertex is a tree of singleton. For each vertex v, we find the trees which can be merged together via adding an edge incident with v. If the degree of v after the merge is at least 3, we merge the trees and update the degrees of the involved vertices. Otherwise we ignore it. It is clear that any vertex of degree two in F is connected to two branching vertices, and therefore the algorithm constructs a maximal leafy forest correctly. By using the union-find data structure, the algorithm runs in $O((m+n)\alpha(m,n))$ time, in which $\alpha(m,n)$ is the inverse Ackermann's function.

The second step of the whole algorithm is to connect all the leafy trees and the singleton trees to form a spanning tree. This can be done in $O(m+n)$ time obviously. Consequently the time complexity in total is also $O((m+n)\alpha(m,n))$, which is almost linear in the size of the graph. The result is stated in the next lemma, and the performance ratio will be shown later.

LEMMA 7.7
Given a graph G, in $O((m+n)\alpha(m,n))$ time, we can construct a spanning tree of G which contains a maximal leafy forest.

7.3.3 Performance ratio

We now turn to the approximation ratio. We are going to establish a bound on the number of leaves of any spanning tree containing a maximal leafy forest.

THEOREM 7.6
Let F be a maximal leafy forest and Y be a spanning tree containing F. Then, for any spanning tree T of G, $|V_1(T)| \leq 3|V_1(Y)|$.

Particularly when T is an MLST, we have that Y is a 3-approximation of an MLST. The following notations are used in the remaining paragraphs of this section:

- $G = (V, E)$: the input graph;
- F: a maximal leafy forest consisting of k leafy trees T_i for $1 \leq i \leq k$;
- Y: a spanning tree containing F;
- T: any spanning tree of graph G.

The following two lemmas are essential to the theorem. Lemma 7.8 shows that the number of leaves in any spanning tree T is upper bounded by $|V(F)|-k+1$, and Lemma 7.9 shows that any spanning tree Y containing a leafy forest F has at least $|V_1(F)| - 2(k-1)$ leaves.

First we give a proof for Theorem 7.6. By Lemma 7.6,

$$|V(T_i)| \leq 3|V_1(T)| - 5.$$

Steiner Trees and Some Other Problems

Since the leafy forest F is the union of the leafy trees, summing up over all leafy trees, we have

$$|V(F)| \le \sum_{i=1}^{k} |V(T_i)| \le \sum_{i=1}^{k}(3|V_1(T)| - 5) = 3|V_1(F)| - 5k. \qquad (7.9)$$

Then, we have

$$\begin{aligned}|V_1(T)| &\le |V(F)| - k + 1 & (\text{ by Lemma 7.8 }) \\ &\le 3|V_1(F)| - 6k + 1 & (\text{ by (7.9) }) \\ &\le 3(|V_1(Y)| + 2(k-1)) - 6k + 1 & (\text{ by Lemma 7.9 }) \\ &\le 3|V_1(Y)|.\end{aligned}$$

It completes the proof. We show the two essential lemmas in the following.

LEMMA 7.8
$|V_1(T)| \le |V(F)| - k + 1$.

PROOF We prove the lemma by establishing two one-to-one functions

$$\begin{aligned}h_1 &: \{T_i | 1 \le i \le k-1\} \longrightarrow V(F) - V_1(T) \\ h_2 &: V_1(T) - V(F) \longrightarrow V(F) - V_1(T)\end{aligned}$$

and showing that the ranges of the two functions are disjoint. As a result,

$$|V(F) - V_1(T)| - |(V_1(T) - V(F))| \ge k - 1,$$

which implies the result in the lemma. Let v_i be a branching vertex in T_i for $1 \le i \le k$. Since T is a spanning tree, there exists a path P_i from v_k to v_i on T for $1 \le i < k$. Define $h_1(T_i)$, $i < k$, to be the first vertex $u_i \in V(T_i)$ while traveling from v_k to v_i along P_i. We need to show that $h_1(T_i)$ is a vertex in F but not a leaf in T and h_1 is one-to-one.

By definition, u_i is adjacent to a vertex not in T_i. By Fact 7.4, u_i is a leaf in T_i. Therefore $u_i \ne v_i$, and it follows that u_i is not a leaf in T since it is an internal vertex of a path in T. Hence $u_i \in V(F) - V_1(T)$. Since the leafy trees T_i are disjoint and $u_i \in V(T_i)$, h_1 is one-to-one.

Now we define the function h_2. Let x_i, $1 \le i \le l$ be the leaf in T but not a vertex in F. Define $h_2(x_i)$ to be the first vertex $w_i \in V(F)$ while traveling along the path from x_i to v_k in T. Once again we need to show that w_i is not a leaf in T and h_2 is one-to-one. Since $w_i \in V(F)$, it belongs to some leafy tree. Also by Fact 7.4, it must be a leaf of the leafy tree. It follows that w_i is not a leaf in T since it is an internal vertex of a path in T. By Fact 7.5, any vertex in F has at most one neighbor not in F. By Fact 7.6, any vertex with two neighbors not in F must have degree two. Therefore $w_i \ne w_j$ for $i \ne j$. Otherwise either w_i has two neighbors not in F or there exists a vertex not in

F but has degree more than 2. Both cases contradict the facts. Consequently h_2 is one-to-one.

We now show that the ranges of the two functions are disjoint. As we have shown, $h_1(T_i) = u_i$ is a leaf in T_i. Assume for a contradiction that $u_i = w_j$ for some j. By the definition of u_i, there exists a path from u_i to v_k passing through only vertices not in T_i. Also, by the definition of w_j, there exists a path from w_j to x_j passing through only vertices not in F. By Fact 7.6, any vertex not in F cannot have degree more than 2. Since x_j is a leaf in T and can not be on the path to v_k, it follows that the only possible intersection of two paths is u_i. It implies that u_i is adjacent to one vertex not in F and to another vertex not is T_i, which contradicts Fact 7.5. □

LEMMA 7.9
$|V_1(Y)| \geq |V_1(F)| - 2(k-1)$.

PROOF Recall that Y is obtained by adding edges to connect the leafy trees and some trees of singleton. Suppose that there are p singleton trees, and p_1 of them are leaves in Y. Clearly $(p+k-1)$ edges are added to connect the trees, which provide an increment $2(p+k-1)$ of the total degree. At least $2(p-p_1) + p_1$ degrees are consumed on the trees of singleton, and therefore the number of destroyed leaves in F is at most

$$2(p+k-1) - (2(p-p_1) + p_1) = 2(k-1) + p_1.$$

Consequently the number of leaves in Y is at least

$$|V_1(F)| - (2(k-1) + p_1) + p_1 = |V_1(F)| - 2(k-1).$$

□

The following theorem summarizes the result of the section.

THEOREM 7.7
Given a graph $G = (V, E)$, a 3-approximation of a maximum leaf spanning tree can be found in $O(\alpha(|V|, |E|)(|V| + |E|))$ time.

7.4 Some Other Problems

We introduce some other important problems in this section. For each of the problems, we give the definition and its current status. Details can be found in the references. The problems are arranged into two parts: network design and computational biology. Many of the problems were also collected in [43] and [4].

7.4.1 Network design

> PROBLEM: Bounded Diameter Spanning Trees
> INSTANCE: Graph $G = (V, E, w)$ and an integer B.
> GOAL: Find a minimum weight spanning tree containing no simple path with more than B edges.

The problem is NP-hard even for any fixed $B \geq 4$ and all edge weights are either 1 or 2 [43]. If all edge weights are equal, the problem is reduced to the minimum diameter spanning tree problem of an unweighted graph and thus can be solved in polynomial time. If $B \leq 3$, the problem can also be solved in polynomial time by checking each edge as the middle edge of the diameter.

> PROBLEM: Capacitated Spanning Trees
> INSTANCE: Graph $G = (V, E)$, source vertex $s \in V$, capacity $c(e) \in Z_0^+$ and length $l(e) \in Z_0^+$ for each edge e, requirement $r(v) \in Z_0^+$ for each nonsource vertex.
> GOAL: Find a minimum weight spanning tree such that the routing load on each tree edge does not exceed the capacity.

The routing load of an edge e is the total requirement of vertices whose path to the source in the tree contains e. The problem is NP-hard; even all requirements are 1 and all capacities are equal to 3, and the geometric version of the problem is also NP-hard [43].

> PROBLEM: Minimum k-Spanning Trees
> INSTANCE: Graph $G = (V, E, w)$ and an integer k.
> GOAL: Find a minimum weight tree spanning at least k vertices.

The problem is NP-hard and can be approximated within ratio 3 [44]. The Euclidean version of the problem admits a PTAS [3].

> PROBLEM: Minimum Degree Spanning Trees
> INSTANCE: Graph $G = (V, E)$.
> GOAL: Find a spanning tree of G such that the maximum degree of the vertices in the tree is minimized.

To determine if a graph has a spanning tree of maximum degree 2 is equivalent to the *Hamiltonian path* problem. Therefore the problem can be easily shown to be NP-hard. But it is approximable with an absolute error guarantee of 1 [39], i.e., there exists a polynomial time algorithm for finding a tree whose degree is at most one more than the optimal. Furthermore the algorithm also works for the generalization of the problem to Steiner trees.

PROBLEM: Minimum Geometric 3-Degree Spanning Trees
INSTANCE: Set S of points in the plane.
GOAL: Find a minimum weight spanning tree for S in which no vertex has degree larger than 3.

The problem admits a PTAS [3]. The 4-degree spanning tree problem also admits a PTAS, but the NP-hardness of the problem is open. The 5-degree problem is polynomial-time solvable. In d-dimensional Euclidean space for $d \geq 3$, the 3-degree spanning tree problem is approximable within 5/3 [65]. The generalization to a metric graph G with a degree bound $\Delta(v)$ for each vertex is called Minimum Bounded Degree Spanning Tree and is approximable within $2 - \min_{v \in V}\{(\Delta(v) - 2)/(\Delta_G(v) - 2) : \Delta_G(v) > 2\}$, where $\Delta_G(v)$ is the degree of v in G [33].

PROBLEM: Minimum Shortest-Paths Trees
INSTANCE: A graph $G = (V, E, w)$ and a vertex r.
GOAL: Find a minimum weight tree T such that $d_T(r, v) = d_G(r, v)$ for each vertex v.

The problem can be solved in $O(|V||E|\log|V|)$ time [51].

7.4.2 Computational biology

In addition to network design, spanning/Steiner tree problems arise in computational biology. In the following, we introduce some problems about evolutionary trees.

An *evolutionary tree*, or *phylogeny*, is a rooted tree structure used to represent the relationship among species in biology, in which each leaf represents one species and each internal node represents the inferred common ancestor of the species in the subtree. A *distance matrix* of n species is a symmetric $n \times n$ matrix M in which $M[i, j]$ is the dissimilarity of species i and j. It is required that $M[i, j] > 0$ for all $0 \leq i, j \leq n$ and $M[i, i] = 0$ for all $0 \leq i \leq n$. Given a distance matrix, the distance-based evolutionary tree reconstruction problem asks for an evolutionary tree T minimizing some objective function subject to some constraints. By different constraints and different objective functions, several problems have been defined and studied. The evolutionary tree reconstruction problem can be thought of as some kind of the Steiner tree problem, in which there are infinite implicit Steiner vertices.

7.4.2.1 Reconstructing trees from perfect data

A matrix M is additive if there exists a tree T realizing M, i.e., $d_T(i, j) = M[i, j]$ for all i and j, and we say that T is an additive tree for matrix M. Note that an additive tree is an unrooted tree. When applied to the evolutionary tree reconstruction problem, the root is determined by some other techniques. For an additive matrix, there are efficient algorithms for constructing the tree realizing the matrix.

PROBLEM: Additive Trees
INSTANCE: A distance matrix M.
GOAL: Find an additive tree for M, or determine that none exists.

The term "additive matrix" is also known as "tree metric" in the literature. The first polynomial time solution was given in [15], where the basic "four-point condition" was proven.

THEOREM 7.8
An $n \times n$ matrix M is additive if and only if, for all $1 \leq i, j, k, l \leq n$,
$$M[i,j] + M[k,l] \leq \max\{M[i,k] + M[j,l], M[i,l] + M[j,k]\}.$$

The theorem implies an $O(n^4)$ time algorithm but several more efficient algorithms were proposed [7, 26, 53, 90]. The most efficient algorithm takes $O(n^2)$ time and the time complexity is optimal.

An ultrametric is a metric with stronger requirement on the distances than a tree metric.

DEFINITION 7.3 An $n \times n$ matrix M is an ultrametric if and only if $M[i,j] \leq \max\{M[i,k], M[j,k]\}$ for all $1 \leq i, j, k \leq n$.

An ultrametric tree is a rooted tree in which the root has the same distance to each of the leaves, i.e., $d_T(u,r) = d_T(v,r)$ for root r of T and all leaves u and v. Let T be an ultrametric tree with root r. It is easy to see that for any internal node v, the subtree T_v rooted at v is also an ultrametric tree.

PROBLEM: Ultrametric Trees
INSTANCE: A distance matrix M.
GOAL: Find an ultrametric tree for M, or determine that none exists.

The next theorem states an important result for the ultrametric.

THEOREM 7.9
A distance matrix is an ultrametric if and only if it can be realized by a unique ultrametric tree.

By the definition of an ultrametric, it is easy to test if a given matrix is an ultrametric in $O(n^3)$ time. The most efficient algorithm to construct an ultrametric tree takes only $O(n^2)$ time (see [49]).

It is important and interesting that a tree metric can be transformed to an ultrametric. Let M be a distance matrix over $\{1..n\}$. Choose any $r \in \{1..n\}$

and any constant $c \geq \max_{ij} M[i,j]$. Define a matrix $M^{(r)}$ with zero diagonal by
$$M^{(r)}[i,j] = c + M[i,j] - M[i,r] - M[j,r]$$
for $i \neq j$. It has been shown that M is a tree metric if and only if $M^{(r)}$ is an ultrametric [7]. The additive tree for M and the ultrametric tree for $M^{(r)}$ have the same topology.

7.4.2.2 Reconstructing tree from nonperfect data

Both additive and ultrametric matrices are idealized data. The discussion of such problems focuses on the theoretic rather than realistic. However, the real data are rarely perfect. It motivates the optimization versions of the evolutionary tree problems. Depending on the desired tree and the objective function, several combinatorial optimization problems were defined and studied. The following is a general definition.

> PROBLEM: Minimum Increment Evolutionary Trees
> INSTANCE: An $n \times n$ distance matrix M.
> GOAL: Find an optimal additive (or ultrametric) tree T such that the leaf set is $\{1..n\}$ and $d_T(i,j) \geq M[i,j]$ for all $1 \leq i,j \leq n$.

The objective function may be one of the following [32]:

- L^1-norm: to minimize $\sum_{ij}(d_T(i,j) - M[i,j])$.
- L^∞-norm: to minimize $\max_{ij}(d_T(i,j) - M[i,j])$.
- tree size: to minimize $w(T)$, i.e., $\sum_{e \in E(T)} w(e)$.

Under L^1-norm, both the additive tree and the ultrametric tree problem have been shown to be NP-hard. The minimum increment ultrametric problem under L^∞-norm can be solved in $O(n^2)$. The minimum size ultrametric tree problem is not only NP-hard but also hard to approximate. There exists an $\varepsilon > 0$ such that the problem cannot be approximated in polynomial time within ratio n^ε unless P=NP. However, if the input is restricted to a metric, the minimum size ultrametric tree problem can be approximated within a ratio $1.5 \log n$ in polynomial time [95]. In [95], a branch and bound algorithm for the problem is also proposed.

> PROBLEM: Evolutionary Tree Insertion with Minimum Increment
> INSTANCE: An additive tree T with leaf set $\{1..n\}$, a species $x \notin \{1..n\}$, and the distance $M[i]$ from x to each species i in T.
> GOAL: Insert x into T optimally such that $d_T(i,x) \geq M[i]$ for all $1 \leq i \leq n$.

Similarly the objective function may be defined by the L^1-norm, L^∞-norm, or tree size. Unlike that most of the tree construction are NP-hard, the insertion problem with any of the above objective functions can be solved in $O(n)$ time [101].

Bibliographic Notes and Further Reading

The study of the Euclidean Steiner minimal tree problem has a long history. Frank Kwang-Ming Hwang and D.S. Richards gave a nice survey up to 1989 [57]. Here we only mention some of the results. The readers can find more details in their survey paper. The MST-based approximation algorithm was proposed by several authors independently, such as [78, 68]. The algorithm we introduced in this chapter is a straightforward implementation. There are more efficient algorithms with the same idea. Basically these algorithms are based on Prim's or Kruskal's MST algorithms without explicitly constructing the metric closure. For example, Y.F. Wu, Peter Widmayer, and C. K. Wong [102] gave a Kruskal-based approach, and H. Takahashi and A. Mastsuyama [84] gave a Prim-based algorithm.

The 11/6-approximation algorithm is due to Alexander Zelikovsky [105], who made the breakthrough of the simple 2-approximation. The result was extended by Piotr Berman and Viswanathan Ramaiyer [11], in which they used 4-tuples to derive a 16/9-approximation algorithm. The Euclidean Steiner ratio is another challenge attracting lots of researchers. E.N. Gilbert and H.O. Pollak [45] conjectured that the ratio is $2/\sqrt{3}$ in the Euclidean plane. Lots of works had been devoted to the improvement of the ratio, and the conjecture was finally proven by Ding-Zhu Du and Frank Kwang-Ming Hwang [31]. For rectilinear distances, Hwang showed that 3/2 is an upper bound of the Steiner ratio [56]. By Zelikovsky's algorithm, the approximation ratio was improved to 11/8, and further to 97/72 by using 4-tuples [11].

The NP-completeness of the decision version of the graph Steiner minimal tree problem was shown by Richard M. Karp [63]. Some restricted versions are also shown to be NP-complete, such as unweighted graphs and bipartite graphs [43]. It was shown [12] that the problem is MAX SNP-hard in the metric that each edge has length either 1 or 2. As a result, it is impossible to derive a polynomial time approximation scheme unless NP=P.

For the minimum diameter spanning tree problem, Jan-Ming Ho et al. [54] studied the geometric version and showed that there is an optimal tree which is either monopolar (a star) or dipolar (a 2-star). Based on the property, they gave an $O(|V|^3)$ time algorithm. The relation between the minimum diameter spanning tree of a graph and its absolute 1-center was pointed out by Refael Hassin and Arie Tamir [52].

The 3-approximation algorithm for the maximum leaf spanning tree is due to Hsueh-I Lu and R. Ravi [73]. In early work of the authors [72], there is another approximation algorithm with the same ratio but much less efficient in time. The MAX SNP-hardness of the problem was shown by Giulia Galbiati, Francesco Maffioli and Angelo Morzenti [41]. They gave an L-reduction from the MINIMUM BOUNDED DOMINATING SET problem to the MLST problem.

For readers interested in computational biology, more problems and details can be found in the book by Dan Gusfield [49].

Exercises

7-1. For any three points in the Euclidean plane, what is the length of the Steiner minimal tree?

7-2. What is the rectilinear distance between two points $(10, 20)$ and $(30, 50)$?

7-3. What are the radius, diameter, and center of a path consisting of n vertices and weighted edges?

7-4. What are the radius, diameter, and center of a cycle consisting of n vertices and weighted edges?

7-5. Is there any graph whose radius is equal to its diameter?

7-6. What are the maximum and the minimum numbers of leaves of an n-vertex tree?

7-7. What is the minimum number of leaves of an n-vertex tree without any vertex of degree two?

7-8. Show that the four-point condition (in Theorem 7.8) entails the triangle inequality and the symmetry of the matrix.

7-9. Sometimes it is assumed that an additive or ultrametric tree is binary. Show that, by inserting zero-length edges, any nonbinary tree can be transformed to a binary tree without changing the distances between leaves.

References

[1] R.K. Ahuja, T.L. Magnanti, and J.B. Orlin. *Network Flows – Theory, Algorithms, and Applications.* Prentice-Hall, 1993.

[2] I. Althöfer, G. Das, D. Dobkin, D. Joseph, and J. Soares. On sparse spanners of weighted graphs. *Discrete Comput. Geom.*, 9:81–100, 1993.

[3] S. Arora. Polynomial time approximation schemes for Euclidean traveling salesman and other geometric problems. *J. ACM*, 45(5):753–782, 1998.

[4] G. Ausiello, P. Crescenzi, G. Gambosi, V. Kann, A. Marchetti-Spaccamela, and M. Protasi. *Complexity and Approximation – Combinatorial Optimization Problems and Their Approximability Properties.* Springer-Verlag, 1999.

[5] B. Awerbuch, A. Baratz, and D. Peleg. Cost-sensitive analysis of communication protocols. In *Proceedings of the 9th Symposium on Principles of Distributed Computing*, pages 177–187, 1990.

[6] V. Bafna, E.L. Lawler, and P. Pevzner. Approximation algorithms for multiple sequence alignment. In *Proceedings of the 5th Combinatorial Pattern Matching conference*, LNCS 807, pages 43–53. Springer-Verlag, 1994.

[7] H.J. Bandelt. Recognition of tree metrics. *SIAM J. Discrete Math*, 3(1):1–6, 1990.

[8] Y. Bartal. Probabilistic approximation of metric spaces and its algorithmic applications. In *Proceedings of the 37th Annual IEEE Symposium on Foundations of Computer Science*, pages 184–193, 1996.

[9] Y. Bartal. On approximating arbitrary metrics by tree metrics. In *Proceedings of the 30th Annual ACM Symposium on Theory of Computing*, pages 161–168, 1998.

[10] R. Bellman. On a routing problem. *Quar. Appl. Math.*, 16:87–90, 1958.

[11] P. Berman and V. Ramaiyer. Improved approximations for the Steiner tree problem. *J. Algorithms*, 17(3):381–408, 1994.

[12] M. Bern and P. Plassmann. The Steiner problem with edge lengths 1 and 2. *Inf. Process. Lett.*, 32(4):171–176, 1989.

[13] K. Bharath-Kumar and J.M. Jaffe. Routing to multiple destinations in computer networks. *IEEE Trans. Commun.*, 31(3):343–351, 1983.

[14] O. Borůvka. O jistém problému minimálním (about a certain minimal problem). *Práca Moravské Přírodovědecké Společnosti*, 3:37–58, 1926. (In Czech.).

[15] P. Buneman. A note on metric properties of trees. *J. Comb. Theory B*, 17:48–50, 1974.

[16] L. Cai. NP-completeness of minimum spanner problems. *Discrete Appl. Math.*, 48:187–194, 1994.

[17] A. Cayley. A theorem on trees. *Quart. J. Math.*, 23:376–378, 1889.

[18] B. Chazelle. A minimum spanning tree algorithm with inverse-Ackermann type complexity. *J. ACM*, 47:1028–1047, 2000.

[19] B. Chazelle. The soft heap: an approximate priority queue with optimal error rate. *J. ACM*, 47:1012–1027, 2000.

[20] D. Cheriton and R. E. Tarjan. Finding minimum spanning trees. *SIAM J. Comput.*, 5:724–742, 1976.

[21] B.V. Cherkassky, A.V. Goldberg, and T. Radzik. Shortest paths algorithms: theory and experimental evaluation. In *Proceedings of the Fifth Annual ACM-SIAM Symposium on Discrete Algorithms*, pages 516–525, 1994.

[22] L.P. Chew. There are planar graphs almost as good as the complete graph. *J. Comput. Syst. Sci.*, 39(2):205–219, 1989.

[23] H.S. Connamacher and A. Proskurowski. The complexity of minimizing certain cost metrics for k-source spanning trees. *Discrete Appl. Math.*, 131:113–127, 2003.

[24] S.A. Cook. The complexity of theorem-proving procedures. In *Proceedings of the 3rd Annual ACM Symposium on Theory of Computing*, pages 151–158, 1971.

[25] T.H. Cormen, C.E. Leiserson, and R.L. Rivest. *Introduction to Algorithms*. The MIT Press, 1994.

[26] J. Culberson and P. Rudnicki. A fast algorithm for constructing trees from distance matrices. *Inf. Process. Lett.*, 30:215–220, 1989.

[27] E.V. Denardo and B.L. Fox. Shortest-route methods: 1. reaching, pruning, and buckets. *Oper. Res.*, 27:161–186, 1979.

[28] E.W. Dijkstra. A note on two problems in connection with graphs. *Numer. Math.*, 1:269–271, 1959.

[29] R. Dionne and M. Florian. Exact and approximate algorithms for optimal network design. *Networks*, 9(1):37–60, 1979.

[30] B. Dixon, M. Rauch, and R. Tarjan. Verification and sensitivity analysis of minimum spanning trees in linear time. *SIAM J. Comput.*, 21:1184–1192, 1992.

[31] D-.Z Du and F.K. Hwang. A proof of the Gilbert-Pollak conjecture on the Steiner ratio. *Algorithmica*, 7:121–135, 1992.

[32] M. Farach, S. Kannan, and T. Warnow. A robust model for finding optimal evolutionary trees. *Algorithmica*, 13:155–179, 1995.

[33] S. Fekete, S. Khuller, M. Klemmstein, B. Raghavachari, and N.E. Young. A network-flow technique for finding low-weight bounded-degree spanning trees. *J. Algorithms*, 24(2):310–324, 1997.

[34] D. Feng and R. Doolittle. Progressive sequence alignment as a prerequisite to correct phylogenetic trees. *J. Mol. Evol.*, 25:351–360, 1987.

[35] M. Fischetti, G. Lancia, and P. Serafini. Exact algorithms for minimum routing cost trees. *Networks*, 39:161–173, 2002.

[36] L.R. Ford, Jr. and D.R. Fulkerson. *Flows in Networks*. Princeton University Press, 1962.

[37] M. Fredman and D. E. Willard. Trans-dichotomous algorithms for minimum spanning trees and shortest paths. *J. Comput. Syst. Sci.*, 48:424–436, 1994.

[38] M.L. Fredman and R.E. Tarjan. Fibonacci heaps and their uses in improved network optimization algorithms. *J. ACM*, 34:596–615, 1987.

[39] M. Fürer and B. Raghavachari. Approximating the minimum-degree Steiner tree to within one of optimal. *J. Algorithms*, 17(3):409–423, 1994.

[40] H. N. Gabow, Z. Galil, T. Spencer, and R. E. Tarjan. Efficient algorithms for finding minimum spanning trees in undirected and directed graphs. *Combinatorica*, 6:109–122, 1986.

[41] G. Galbiati, F. Maffioli, and A. Morzenti. A short note on the approximability of the Maximum Leaves Spanning Tree Problem. *Inf. Process. Lett.*, 52(1):45–49, 1994.

[42] G. Gallo and S. Pallottino. Shortest paths algorithms. *Ann. Oper. Res.*, 13:3–79.

[43] M.R. Garey and D.S. Johnson. *Computers and Intractability: A Guide to the Theory of NP-Completeness*. W.H. Freeman and Company, San Francisco, 1979.

References

[44] N. Garg. A 3-approximation for the minimum tree spanning k vertices. In *Proceedings of the 37th Annual IEEE Symposium on Foundations of Computer Science*, pages 302–309, Burlington, Vermont, 1996.

[45] E.N. Gilbert and H.O. Pollak. Steiner minimal trees. *SIAM J. Appl. Math.*, 16(1):1–29, 1968.

[46] A.V. Goldberg. A simple shortest path algorithm with linear average time. In *Proceedings of the 9th Annual European Symposium Algorithms*, pages 230–241, 2001.

[47] R. L. Graham and P. Hell. On the history of the minimum spanning tree problem. *Ann. Hist. Comput.*, 7:43–57, 1985.

[48] D. Gusfield. Efficient methods for multiple sequence alignment with guaranteed error bounds. *Bull. Math. Biol.*, 55:141–154, 1993.

[49] D. Gusfield. *Algorithms on Strings, Trees, and Sequences – Computer Science and Computational Biology*. Cambridge University Press, 1997.

[50] T. Hagerup. Improved shortest paths in the word RAM. In *Proceedings of the 27th International Colloquium on Automata, Languages and Programming*, pages 61–72, 2000.

[51] P. Hansen and M. Zheng. Shortest shortest path trees of a network. *Discrete Appl. Math.*, 65:275–284, 1996.

[52] R. Hassin and A. Tamir. On the minimum diameter spanning tree problem. *Inf. Process. Lett.*, 53:109–111, 1995.

[53] J. Hein. An optimal algorithm to reconstruct trees from additive distance data. *Bull. Math. Biol.*, 51:597–603, 1989.

[54] J.-M. Ho, D.T. Lee, C.-H. Chang, and C.K. Wong. Minimum diameter spanning trees and related problems. *SIAM J. Comput.*, 20:987–997, 1991.

[55] T.C. Hu. Optimum communication spanning trees. *SIAM J. Comput.*, 3:188–195, 1974.

[56] F.K. Hwang. On Steiner minimal trees with rectilinear distance. *SIAM J. Appl. Math.*, 30:104–114, 1976.

[57] F.K. Hwang and D.S. Richards. Steiner tree problems. *Networks*, 22:55–89, 1992.

[58] O.H. Ibarra and C.E. Kim. Fast approximation algorithms for the knapsack and sum of subset problems. *J. ACM*, 22:463–468, 1975.

[59] V. Jarník. O jistém problému minimálním (about a certain minimal problem). *ráca Moravské Přírodovědecké Společnosti*, 6:57–63, 1930.

[60] D.S. Johnson, J.K. Lenstra, and A.H.G. Rinnooy Kan. The complexity of the network design problem. *Networks*, 8:279–285, 1978.

[61] D. R. Karger, P. N. Klein, and R. E. Tarjan. A randomized linear-time algorithm to find minimum spanning trees. *J. ACM*, 42:321–328, 1995.

[62] O. Kariv and S.L. Hakimi. An algorithmic approach to network location problems I: The p-centers. *SIAM J. Appl. Math.*, 37:513–537, 1979.

[63] R.M. Karp. Reducibility among combinatorial problems. In R.E. Miller and J.W. Thatcher, editors, *Complexity of Computer Computations*, pages 85–103. Plenum Press, New York, 1975.

[64] S. Khuller, B. Raghavachari, and N. Young. Balancing minimum spanning trees and shortest-path trees. *Algorithmica*, 14:305–321, 1995.

[65] S. Khuller, B. Raghavachari, and N. Young. Low degree spanning trees of small weight. *SIAM J. Comput.*, 25:355–368, 1996.

[66] V. King. A simpler minimum spanning tree verification algorithm. *Algorithmica*, 18:263–270, 1997.

[67] J. Komlós. Linear verification for spanning trees. *Combinatorica*, 5:57–65, 1985.

[68] L. Kou, G. Markowsky, and L. Berman. A fast algorithm for Steiner trees. *Acta Inform.*, 15(2):141–145, 1981.

[69] J. B. Kruskal. On the shortest spanning subtree of a graph and the travelling salesman problem. *Proc. Amer. Math. Soc.*, 7:48–50, 1956.

[70] E.L. Lawler. Fast approximation algorithms for knapsack problems. *Math. Oper. Res.*, 4(4):339–356, 1979.

[71] C. Levcopoulos and A. Lingas. There are planar graphs almost as good as the complete graphs and as cheap as minimum spanning trees. *Algorithmica*, 8(3):251–256, 1992.

[72] H.-I Lu and R. Ravi. The power of local optimization: Approximation for maximum-leaf spanning trees. In *Proceedings of the 13th Allerton Conference on Communication, Control and Computing*, pages 533–542, 1992.

[73] H.-I Lu and R. Ravi. Approximating maximum leaf spanning trees in almost linear time. *J. Algorithms*, 29(1):132–141, 1998.

[74] U. Meyer. Single-source shortest paths on arbitrary directed graphs inlinear average time. In *Proceedings of the 12th Annual ACM-SIAM Symposium on Discrete Algorithms*, pages 797–806, 2001.

[75] D. Peleg and J.D. Ullman. An optimal synchronizer for the hypercube. In *Proceedings of the 6th Symposium on Principles of Distributed Computing*, pages 77–85, 1987.

[76] S. Pettie and V. Ramachandran. An optimal minimum spanning tree algorithm. *J. ACM*, 49:16–34, 2002.

[77] P. Pevzner. Multiple alignment, communication cost, and graph matching. *SIAM J. Appl. Math.*, 52:1763–1779, 1992.

[78] J. Plesnik. The complexity of designing a network with minimum diameter. *Networks*, 11:77–85, 1981.

[79] R. C. Prim. Shortest connection networks and some generalizations. *Bell. Syst. Tech. J.*, 36:1389–1401, 1957.

[80] H. Prüfer. Never beweis eines satzes über permutationen. *Arch. Math. Phys. Sci.*, 27:742–744, 1918.

[81] R. Raman. Priority queues: small, monotone and trans-dichotomous. In *Proceedings of the 4th Annual European Symposium Algorithms*, pages 121–137, 1996.

[82] R. Raman. Recent results on single-source shortest paths problem. *SIGACT News*, 28:81–87, 1997.

[83] D. Sankoff and J. B. Kruskal, editors. *Time Warps, String Edits and Macromolecules: The Theory and Practice of Sequence Comparison*. Addison Wesley, 1983.

[84] H. Takahashi and A. Mastsuyama. An approximate solution for the Steiner problem in graphs. *Math. Jap.*, 24:573–577, 1980.

[85] R. E. Tarjan. Efficiency of a good but not linear set-union algorithm. *J. ACM*, 22:215–225, 1975.

[86] R. E. Tarjan. Applications of path compressions on balanced trees. *J. ACM*, 26:690–715, 1979.

[87] M. Thorup. Undirected single-source shortest paths with positive integer weights in linear time. *J. ACM*, 46:362–394, 1999.

[88] M. Thorup. On RAM priority queues. *SIAM J. Comput.*, 30:86–109, 2000.

[89] P.M. Vaidya. A sparse graph almost as good as the complete graph on points in k dimensions. *Discrete Comput. Geom.*, 6:369–381, 1991.

[90] M. Waterman, T. Smith, M. Singh, and W. Beyer. Additive evolutionary trees. *J. Theor. Biol.*, 64:199–213, 1977.

[91] M.S. Waterman. *Introduction to Computational Biology*. Chapman & Hall, CRC Press, 1995.

[92] R. Wong. Worst-case analysis of network design problem heuristics. *SIAM J. Algebra. Discr.*, 1:51–63, 1980.

[93] B.Y. Wu. A polynomial time approximation scheme for the two-source minimum routing cost spanning trees. *J. Algorithms*, 44:359–378, 2002.

[94] B.Y. Wu. Approximation algorithms for optimal p-source communication spanning trees. Unpublished manuscript, 2003.

[95] B.Y. Wu, K.-M. Chao, and C.Y. Tang. Approximation and exact algorithms for constructing minimum ultrametric trees from distance matrices. *J. Comb. Optim.*, 3:199–211, 1999.

[96] B.Y. Wu, K.-M. Chao, and C.Y. Tang. Approximation algorithms for some optimum communication spanning tree problems. *Discrete Appl. Math.*, 102:245–266, 2000.

[97] B.Y. Wu, K.-M. Chao, and C.Y. Tang. Approximation algorithms for the shortest total path length spanning tree problem. *Discrete Appl. Math.*, 105:273–289, 2000.

[98] B.Y. Wu, K.-M. Chao, and C.Y. Tang. A polynomial time approximation scheme for optimal product-requirement communication spanning trees. *J. Algorithms*, 36:182–204, 2000.

[99] B.Y. Wu, K.-M. Chao, and C.Y. Tang. Light graphs with small routing cost. *Networks*, 39:130–138, 2002.

[100] B.Y. Wu, G. Lancia, V. Bafna, K.-M. Chao, R. Ravi, and C.Y. Tang. A polynomial time approximation scheme for minimum routing cost spanning trees. *SIAM J. Comput.*, 29:761–778, 2000.

[101] B.Y. Wu and C.Y. Tang. An $O(n)$ algorithm for finding an optimal position with relative distances in an evolutionary tree. *Inf. Process. Lett.*, 63:263–269, 1997.

[102] Y.F. Wu, P. Widmayer, and C.K. Wong. A faster approximation algorithm for the Steiner problem in graphs. *Acta Inform.*, 23(2):223–229, 1986.

[103] Y. Xu, V. Olman, and D. Xu. Clustering gene expression data using a graph-theoretic approach: An application of minimum spanning trees. *Bioinformatics*, 18:536–545, 2002.

[104] A. Yao. An $O(|E|\log\log|V|)$ algorithm for finding minimum spanning trees. *Inf. Process. Lett.*, 4:21–23, 1975.

[105] A. Zelikovsky. An 11/6-approximation algorithm for the network Steiner problem. *Algorithmica*, 9:463–470, 1993.

Index

δ-path, 67
δ-spine, 67

absolute 1-center, 161
additive tree, 171
alignment, 80
    multiple sequence, 80–82
    sum-of-pair, 80, 124
    tree-driven, 81
assignment problem, 75

Bellman-Ford algorithm, 33
Borůvka's algorithm, 11
bounded diameter spanning tree, 169
branch, 48

CAL, see cut and leaf set
capacitated spanning tree, 169
Cayley's formula, 1
center, 154
centroid, 46, 88
clustering gene expression data, 17
cut and leaf set, 67

diameter, 154
Dijkstra's algorithm, 25
distance matrix, 170

eccentricity, 154
Eulerian cycle, 131
Eulerian graph, 131
evolutionary tree, 170
evolutionary tree insertion problem, 172
exact cover by 3-sets, 148

four-point condition, 171

Hamiltonian cycle, 150
Hamiltonian path, 169

knapsack problem, 126

Kruskal's algorithm, 15

LART, see light approximate routing cost spanning tree
LASF, see light approximated shortest-path forest
LAST, see light approximate shortest-paths tree
LCS, see longest common subsequence
leafy forest, 164
leafy tree, 163
light approximate routing cost spanning tree, 130
light approximate shortest-paths tree, 129
light approximated shortest-path forest, 137
longest common subsequence, 84

maximum leaf spanning tree, 162
MDST, see minimum diameter spanning tree
median, 45
metric closure, 58
metric graph, 57
minimum k-spanning tree, 169
minimum bounded degree spanning tree, 170
minimum cut, 94
minimum degree spanning tree, 169
minimum diameter spanning tree, 157
minimum geometric 3-degree spanning tree, 170
minimum increment evolutionary tree, 172
minimum routing cost spanning tree, 41, 85, 129
    p-source, 86
    Steiner, 127
minimum routing cost spanning trees
    p-source, 109

minimum shortest-paths tree, 170
minimum spanning tree, 9, 19, 129, 148
MLST, *see* maximum leaf spanning tree
MRCT, *see* minimum routing cost spanning tree
ΔMRCT, 61
MST, *see* minimum spanning tree

OCT, *see* optimal communication spanning tree
optimal communication spanning tree, 85
    p-source, 85
optimal product-requirement communication spanning tree, 85
optimal sum-requirement communication spanning tree, 85, 104

phylogeny, 170
Prüfer sequence, 2
Prim's algorithm, 13
PROCT, *see* optimal product-requirement communication spanning tree

radius, 154
rectilinear, 148
routing cost, 41
routing load, 41
    product-requirement, 89
    sum-requirement, 104

satisfiability problem, 110
scaling and rounding, 101, 126
separator, 48
    minimal, 48
    path, 53
shortest total path length spanning tree, 83
shortest-paths tree, 23, 44
SMT, *see* Steiner tree, minimal
solution decomposition, 46
SP-alignment, *see* sum-of-pair, alignment
spanner, 145
spanning tree, 1
    counting, 1

    minimum, *see* minimum spanning tree
SPT, *see* shortest-paths tree
SROCT, *see* optimal sum-requirement communication spanning tree
star, 50
    k-star, 62, 67
    configuration, 74
    general, 50
Steiner ratio, 140, 151
Steiner tree, 147–154, 170
    Euclidean, 148
    graph, 148
    minimal, 140, 148

traveling salesperson problem, 18, 150
tree metric, 171
TSP, *see* traveling salesperson problem

ultrametric, 171
ultrametric tree, 171